FIRE FIGHTING PUMPING SYSTEMS AT INDUSTRIAL FACILITIES

FIRE FIGHTING PUMPING SYSTEMS AT INDUSTRIAL FACILITIES

Second Edition

by

Dennis P. Nolan, P.E., PhD.

ELSEVIER

Amsterdam • Boston • Heidelberg • London • New York • Oxford
Paris • San Diego • San Francisco • Singapore • Sydney • Tokyo
Academic Press is an imprint of Elsevier

ACADEMIC PRESS

Gulf Professional Publishing is an imprint of Elsevier
The Boulevard, Langford Lane, Kidlington, Oxford OX5 1GB, UK
30 Corporate Drive, Suite 400, Burlington, MA 01803, USA

Second edition 2011

British Library Cataloguing Number: 628.9'25-dc22

Library of Congress Control Number: 2011925951

ISBN: 978-0-12-810378-4

For information on all Elsevier publications visit our website at elsevierdirect.com

Dedicated to
Kushal, Nicholas and Zebulon

CONTENTS

ABOUT THE AUTHOR

Dennis P. Nolan has had a long career devoted to risk engineering, fire protection engineering, loss prevention engineering and system safety engineering. He holds a Doctor of Philosophy degree in Business Administration from Berne University, a Master of Science degree in Systems Management from Florida Institute of Technology and a Bachelor of Science degree in Fire Protection Engineering from the University of Maryland. He is also a registered Professional Engineer in Fire Protection Engineering in the State of California.

He is currently associated with the Fire Prevention Engineering staff of the Saudi Arabian Oil Company (Saudi Aramco), located in Abqaiq, Saudi Arabia, which contains the largest oil and gas facilities in the world. The magnitude of the risks, worldwide sensitivity and foreign location make this one of the most highly critical Highly Protected Risk (HPR) operations in the world. He has also been associated with Boeing, Lockheed, Marathon Oil Company and Occidental Petroleum Corporation in various fire protection engineering, risk and safety roles. These positions were in several locations in the United States and overseas. As part of his career, he has examined oil production, refining and marketing facilities in various severe and unique worldwide locations, including Africa, Asia, Europe, the Middle East, Russia and North and South America. His activity in the aerospace field has included engineering support for the NASA Space Shuttle launch facilities at Kennedy Space Center (and those undertaken at Vandenburg Air Force Base, California) and classified "Star Wars" defense systems.

Dr Nolan has received numerous safety awards and is a member of the American Society of Safety Engineers, and previously the National Fire Protection Association, Society of Petroleum Engineers, and the Society of Fire Protection Engineers. He was also a member of the Fire Protection Working Group of the UK Offshore Operators Association (UKOOA). He is the author of many technical papers and has written several previous books, and professional articles in various international fire safety publications. He has also written several other books which include: *Application of HAZOP and What-If Safety Reviews to the Petroleum, Petrochemical and Chemical Industries*; *Handbook of Fire and Explosion Protection Engineering Principles for Oil, Gas, Chemical and Related Facilities*; *Encyclopedia of Fire Protection*; *Safety and Security Review for the Process Industries: Application of HAZOP, PHA and What-If Reviews*; and *Loss Prevention and Safety Control: Terms and Definitions.* Dr Nolan has also been listed for many years in *Who's Who in California,* has been included in the 16th Edition of *Who's Who in the World* and listed in "Living Legends" (2004), published by the International Biographical Center, Cambridge, England.

ACKNOWLEDGMENTS

Grateful acknowledgment is due to the following people, for their assistance, or for the use of their material in this book:

Steven Chaloupka, Amarillo Gear Company; Sandro Paliotti, Armstrong Darling, Inc.; Aurora Pump Company; John McConnell, Chemguard, Inc.; Louis Sotis and Pamela J. Munslow, Factory Mutual; R.V. Stanford, Lloyd's Register of Shipping; Warren E. Hill III, Metron, Inc.; Geir Christian Helgesen, Marketing and Communications Manager, Frank Mohn Flatoy AS; Dennis J. Berry, National Fire Protection Association; Howard Collins, Occidental Petroleum Corporation; Brian Henry, Patterson Pump Company.

NOTICE

PREFACE

This book describes fixed firewater pump installations for industrial facilities from the viewpoint of the end users, fire protection engineers, loss prevention professionals and those just entering a career in which decisions about fire pump installations must be made. Therefore, much background information is given for the necessary requirements and usefulness of a firewater pump and the services that interface with it.

It is assumed that the reader is to some extent generally knowledgeable about hydraulics for pumps and pumping systems, therefore this book is not concerned with those detailed design aspects. Many excellent reference books are available which provide adequate guidance in the design of firewater installations to calculate water flow, pressure and other hydraulic features and concerns associated with fire pump installations.

This book's primary objective is the provision of practical information and basic background design principles on the application of fixed pumps for fire-fighting purposes at industrial facilities, both onshore and offshore. Where specific details are necessary and pertinent to the discussion they are provided, otherwise, these can be found from the applicable fire codes and engineering practices to be applied to the facility. Experience from the installation of fire pumps in the petroleum and chemical industries, historical data, manufacturers specification sheets and regulatory code requirements have been drawn upon for the preparation of the information in this book.

All fire water pump installations should meet the requirements of local ordinances and applicable fire codes for the facility. This book does not intend to replace or supplement the legal requirements of those documents and the required responsibility to meet them rest with the owner of the facility.

INTRODUCTION

Millions of unexpected fires occur every year and cause damage which amounts to several billions of dollars. Fortunately, most fires are small and easily suppressed. Water is the universal agent to control and suppress unwanted fires, be it a small incident or a large industrial conflagration. Water suppresses fires by oxygen deprivation through smothering and by heat absorption. 3.8 liters (one gallon) of water will absorb about 1,512 k cal (6,000 Btu's) when vaporized to steam in a fire. It is the most efficient, economical and most readily available medium for extinguishing fires of a general nature.

Firewater pumps are used to raise, transfer or increase the pressure or quantities of water applied in fire fighting. Therefore, at industrial facilities a firewater pump is commonly employed to effectively supply and deliver fire-fighting water. Industrial firewater pumps are normally of centrifugal design and this book is mainly concerned with these types. Individually they can range in size from 95 l/min (25 gpm) to as much as 47,332 l/min (12,500 gpm). Industrial facilities contain hazards that are unique to their own operations. These hazards vary with the type of processes, structures and materials handled. Standard and sometimes unique firewater pump installations are therefore provided to meet these risks within these industries.

Centrifugal pumps are commonly employed in the provision of firewater supplies. The basic centrifugal pump consists of a rotating disk molded into vanes referred to as the impeller. The impeller is encased in a housing to channel and direct the produced liquid flow. Water enters near the center of the impeller; whereby motion is imparted to the fluid through the rotation of the vanes and the water is then discharged though the outlet of the casing. By varying the particular designs and arrangements of centrifugal pumps, they can be constructed to suit specific needs or requirements.

The study of fluids in motion is called fluid dynamics. Pumps provide for the movement of fluids and are therefore associated with the principles of fluid dynamics. The design of pumps to achieve proper fluid dynamic principles is beyond the scope of this book but can be found in other excellent reference books on the subject of pump design.

One of the main reasons for writing this book is to provide an adequate reference book from the perspective of industrial users (rather than from regulatory enforcement, i.e. National Fire Protection Association (NFPA) 20, LPC, S.I. 611, etc.). One can learn from this book the general and specific installation arrangements for fire pumps. These concepts can be easily understood and referred to later. Previously, a dedicated book on the installation and features of firewater pumps utilized for high-risk industrial exposures was generally unavailable. Specific prescriptive requirements from fire codes are mentioned for any industrial firewater pump design or purchase and a brief mention is provided in reference books on pumps or fire prevention practices. This is commonly thought of as all that is necessary for a fire water pump installation. Various pump designs, options and features are available to the designer at many economic levels. Also, since a pump is only one portion of a pumping system, all components and supporting services need to be examined to fulfill the review of firewater pumping needs.

The first reference to fire pumps in NFPA fire codes was in 1896 and was primarily directed toward a source of backup water supply to sprinkler systems, standpipes and hydrants. Since then, firewater pumps have been considered a prime source for the supply of fire-fighting water in the industrial world. Nowadays, almost all industrial facilities are routinely provided with firewater pumps. In some cases, the provision of the firewater pumping system may account for a considerable economic percentage of the overall safety features provided to a facility. This merits a critical examination of the design to ensure a cost effective, reliable and efficient firewater pumping installation is achieved.

In the investigation of fire incidents, the performance of the facility firewater pumps is usually one of the first issues raised by the investigators. Additionally, insurance underwriters are also keenly interested in the installation and specific data of the annual performance testing of fire pumps to demonstrate that requirements are being met during their surveys of the facility and initial assessment of its protection measures. It has been stated on one occasion that the failure of the firewater system in 12 of the 100 worst industrial fire incidents has been a major contributing factor in the resulting large scale damage that ensued. Roughly stated, approximately 10 percent of all industrial fire incidents involve failure of the firewater system to meet its objective requirements. Thus, it is imperative that these systems be designed and installed to provide reliable and high integrity service. It is most probable that the 12 failed firewater systems mentioned above were all in compliance with local and national codes for the firewater system, yet they still failed to give adequate service to the incident.

Not every pump that is manufactured is "permitted" to be used in a firewater pumping system. The major difference between the wide range of commercially available pumps and *qualified* fire pumps are the requirements (i.e. standards and specifications) for their manufacture and installation established by Underwriters Laboratories (UL), Factory Mutual (FM) and NFPA or other national and international regulatory approval agencies. Yet, many industrial locations may incidentally use "ordinary" pumps that have not been listed or approved for firewater service.

Some pumping systems are available that may meet the letter or intent of these requirements but have not been officially submitted for approval or listing marks, although they "may" be perfectly satisfactory for firewater service. Their features and options must be evaluated by the authority having jurisdiction, to achieve an acceptable, practical, reliable and economical firewater pump installation.

The installation of a firewater pump appears to be a fairly simple task. Simple errors, oversights, complacency or even economic pressures may lead to a major impact on a facility during an incident because some critical feature of the firewater pumping system was not provided, maintained or was overlooked. This book hopes to provide some insight into the typical arrangements and features that should be considered during the design, installation, operation, maintenance, and testing of firewater pumps for industrial facilities to avoid these circumstances.

Numerous other references are provided at the end of the book to further assist the reader in the installation and maintenance of firewater pumps.

LIST OF TABLES

LIST OF FIGURES

LIST OF ACRONYMS

2oo3	Two Out Of Three.
A60	Firewall rating able to withstand a cellulosic fire for 60 minutes with average surface temperature on the unexposed side of 139 °C (282 °F) above the original temperature.
AC	Alternating Current.
AFFF	Aqueous Film Forming Foam.
AGMA	American Gear Manufacturers' Association.
AHJ	Authority Having Jurisdiction.
ANSI	American National Standards Institute.
AODC	Association of Offshore Diving Contractors.
API	American Petroleum Institute.
ARV	Air Release Valve.
ASTM	American Society for Testing of Materials.
ATS	Automatic Transfer Switch.
AWWA	American Water Works Association.
BHP	Brake Horsepower.
BLEVE	Boiling Liquid Expanding Vapor Explosion
BS	British Standard.
CA	Certifying Authority.
CCW	Counter-Clockwise.
CEA	Comite Europeen des Assurances.
CFM	Cubic Feet per Minute.
CO_2	Carbon Dioxide.
CPVC	Chlorinated Polyvinyl Chloride.
CRV	Circulation Relief Valve.
CV	Check Valve.
CW	Clockwise.
Cu-Ni	Copper-Nickel.
DC	Direct Current.
DCS	Distributed Control System.
DIN	German Standards.
EEMUA	Engineering Equipment and Materials Users Association.
EPSS	Emergency Power Supply System.
ESD	Emergency Shutdown System.
EIV	Emergency Isolation Valve.
FAT	Factory Acceptance Test.
FCV	Flow Control Valve.

FI	Formal Interpretation.
FMEA	Failure Mode and Effects Analysis.
FM	Factory Mutual.
FMRC	Factory Mutual Research Corporation.
FP	Fire Pump.
FPA	Fire Protection Association (UK).
FPE	Fire Protection Engineer.
fps	feet per second.
FRA	Firewater Reliability Analysis.
FTA	Fault Tree Analysis.
gpm	gallons per minute.
GRP	Glass Reinforced Plastic.
GS	Gravity Sewer.
H2S	Hydrogen Sulfide.
HAZOP	Hazard and Operability Study.
HP	Horsepower.
HPR	Highly Protected Risk.
HSE	Health and Safety Executive (UK).
HSI	Hydraulics Standards Institute.
HVAC	Heating, Ventilation and Air Conditioning.
IBC	International Building Code.
ID	Internal Diameter.
IES	Illuminating Engineering Society.
IMO	International Maritime Organization.
IP	Institute of Petroleum (UK).
IRI	Industrial Risk Insurers.
ISO	International Organization for Standardization.
ISO	Insurance Services Office.
kPa	kilopascals.
kW	kilowatt.
LAT	Lowest Astronomical Tide.
LEL	Lower Explosive Limit.
LPC	Loss Prevention Council (UK).
LPCB	Loss Prevention Certification Board (UK).
LPG	Liquefied Petroleum Gas.
LTA	Less Than Adequate.
l/min	Liters per minute.
m	meters.
mg/L	milligrams/liter.
MMS	Minerals Management Service.
MTBF	Mean Time Between Failures.
MODU	Mobile Offshore Drilling Unit.
NEC	National Electrical Code.
NEMA	National Electrical Manufacturers Association.

NFPA	National Fire Protection Association.
NFAC	National Fire Alarm Code.
NPSH	Net Positive Suction Head.
NPSHa	Net Positive Suction Head Available.
NPSHr	Net Positive Suction Head Required.
NRV	Non-Return Valve (i.e. Check Valve).
OCMA	Oil Company Materials Association.
OD	Outside Diameter.
OSHA	Occupational Safety and Health Administration.
OS&Y	Outside Stem and Yoke.
PCV	Pressure Control Valve.
pH	Relative Ranking of Water Conditions, Acidity to Alkalinity (0 to 14).
PI	Pressure Indicator.
P & ID	Piping and Instrumentation Diagram.
PIV	Post Indicator Valve.
PLC	Programmable Logic Controller.
ppm	parts per million.
PS	Pressure Switch.
psi	pounds per square inch.
psia	Pounds per square inch, absolute.
psig	Pounds per square inch, gage.
PSSR	Pre-Startup Safety Review.
PSV	Pressure Safety Valve.
PTFE	Polytetrafluroethylene (Teflon).
PVC	Polyvinyl Chloride.
QA	Quality Assurance.
QC	Quality Control.
RP	Recommended Practice.
rpm	revolutions per minute.
RTR	Reinforced Thermosetting Resin.
RV	Relief Valve.
SAT	Site Acceptance Test.
SG	Special Grade.
SFPE	Society of Fire Protection Engineers.
SI	Statutory Instrument (UK).
SOLAS	Safety of Life at Sea.
SPF	Single Point Failure.
TDS	Total Dissolved Solids.
TVA	Torsional Vibration Analysis.
UBC	Uniform Building Code.
UFC	Uniform Fire Code.
UK	United Kingdom.
UKOOA	UK Offshore Operators Association.
UVCE	Unconfined Vapor Cloud Explosion.

UL	Underwriters' Laboratories, Inc.
ULC	Underwriters' Laboratories, Inc, Canada.
US	United States.
WCCE	Worst Case Creditable Event.

1 HISTORICAL APPLICATIONS OF FIREWATER PUMPING SYSTEMS

1.1. INTRODUCTION

A pump is a device that utilizes energy to raise, transport or compress fluids and gases. The term pump is used for liquid handling devices, whereas a compressor is used when the pressure of a gas is increased. The term "fire engine" was classically referred to any device that was used to extinguish fires. Current English language linguistics refer to a fire engine as a mobile fire apparatus (i.e., pumper), while firewater pumping systems are commonly referred to when fixed installations are involved.

Pumping devices have been in use for thousands of years and applied to a variety of uses. Most of the technological improvements made in water pumping systems have occurred within the last 100 years. The version of the pump that is commonly employed today for firewater service is the centrifugal pump, which was invented during the Industrial Revolution of the 1800s and is now almost universally adopted. Prior to this, reciprocating or rotary water pumps were used which were operated by hand, wind, or steam power.

1.2. ANCIENT WATER PUMPS

The first type of "pump" was probably used by the ancient Egyptians sometime around 2,000 BC. They used waterwheels with buckets to provide for agricultural irrigation. In the third century BC, Ctesibius of Alexandria invented a water pump for fire extinguishment. Apparently, Alexandria had some type of hand-operated fire engine, similar to those used in Europe and America in the eighteenth century. Subsequently, at around 200 BC, the Greeks invented a reciprocating pump.

In the first century BC, Heron of Alexandria was credited with producing an improved type of reciprocating fire pump based on the pump invented by Ctesibius. This pump was essentially a suction lift pump, but modified to a cylinder force pump. The pump had two pistons, each within its own cylinder that had a foot valve. The pistons were connected by a rocker arm that pivoted on a center post. The cylinders were supplied with water through a foot valve located at the bottom of the cylinder. By lifting and forcing the pistons down with the rocker arm, water was lifted and force was applied so that it could be "pushed" out of a nozzle connected to the top of the cylinder. The nozzle was mounted so that it could pivot and swivel in any direction. This allowed for water application on a nearby fire incident (see Figure 1-1). Piston pumps were also reportedly used as flame throwers which Greek ships used as weapons, which probably

Fire Fighting Pumping Systems at Industrial Facilities. DOI: 10.1016/B978-1-4377-4471-2.00001-3

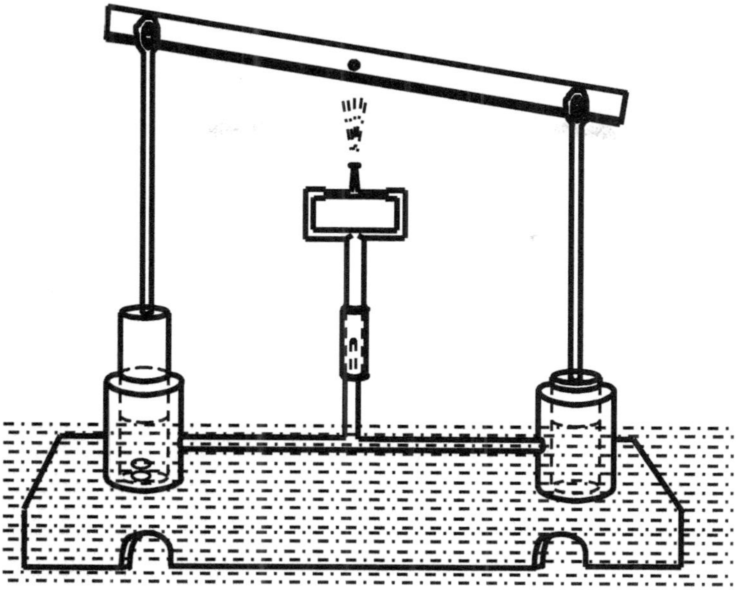

Figure 1-1 Ancient Pump of Antiquity

used a petroleum-based liquid that was ignited. Pliny (23-79 AD) also mentions the use of "fire engines" in ancient Rome for fire-fighting purposes.

With the fall of the Roman Empire, large cities disappeared in the West and therefore the simultaneous destruction of a large number of buildings by large fires did not occur. The development of a fire pump was therefore not in demand. When larger cities again appeared in the Middle Ages, the destruction of cities by conflagration resumed. It was not until the end of the fifteenth century that the reciprocating fire pump was re-invented. The rapid industrialization of the seventeenth, eighteenth and nineteenth centuries, and the ensuing, frequent conflagrations of large cities, saw the development of many types and applications of pumps and water distribution systems specifically for fire-fighting.

1.3. RECIPROCATING HAND AND STEAM-DRIVEN FIRE PUMPS

The reciprocating water pump remained in service until late in the Industrial Revolution. The main reason for this was the lack of a high power source. Most industrial energy sources at that time were of approximately 7.5 kilowatts (10 horsepower) or less capacity (i.e. windmills, waterwheels, animal and human efforts, etc.). Without a sufficient power source to rapidly move water supplies, only limited capacities could be achieved.

The fire pump of this time was commonly mounted on a cart or carriage and brought to the scene of a fire by a team of horses. A tub or reservoir of water was provided on the carriage at the base of the pump. This reservoir was filled by means of a bucket brigade by the local populace. Later, with the provision of street water mains, fire engines connected directly to fire hydrants. This type of mobile fire pump was used and improved upon until the late 1800s. When steam power was developed, it was applied to drive the reciprocating firewater pump in lieu of men.

Reciprocating hand pumps for supplying water to extinguish fires (or to pump bilge water/wash the decks) were also an essential part of the fittings available to late eighteenth-century English ships (i.e. c. 1772). The first fireboats in the United States appeared in 1800 for New York City. They used a hand-operated pump and were imported from England at a cost of, at the time, $4,000 each.

The first fire engine made in America was built for the city of Boston. It was made in 1654 by Joseph Jencks, an iron maker of Lynn, Massachusetts, and was operated by relays of men using handles. The production of this fire engine was the result of a disastrous fire suffered by the city in January 1653. By 1715, Boston had six fire companies with engines of English manufacture. The steam-pump fire engine was actually introduced in London in 1829 by John Ericsson and John Braithwait. It was in use in many large cities by the 1850s. Most steam pumpers were equipped with reciprocating piston pumps, although a few rotary pumps were also used. Some were self-propelled, but most used horses for propulsion, conserving the steam pressure for the pump. The first *practical* fire engine in America was the "Uncle Joe Ross", invented by Alexander Bonner Latta. It was constructed in 1852 in Cincinnati, Ohio. It weighed approximately four tons and required four horses to pull it, and used its own power. It could provide up to six streams of water. A single stream had a 4.4 cm (1 ¾ inch) diameter, and it had a reach of 73 meters (240 ft.). The first steam fire engine in America was actually designed and built in 1841 by Paul R. Hodge. It was 4.3 meters (14 ft) long and weighed 7257 kgs (eight tons). Because of its weight and the sparks produced from its stack, it was later abandoned. A steam fire engine used by the New York Fire Department remained in service as late as 1932.

1.4. ROTARY PUMPS

An early centrifugal type rotary pump was made in the early seventeenth century. It could pump water about nine meters (30 ft). A more effective rotary pump was made by a Frenchman named Dietz in the late nineteenth century. A pump similar to Dietz's was shown at the London Great Exhibition of 1851 and received wide acclaim.

1.5. INVENTION OF THE CENTRIFUGAL PUMP

The true centrifugal pump was not developed until late in the 1600s. Denis Papin (1647-c. 1712), a French physicist and inventor, produced a centrifugal pump with straight vanes. In 1851, John G. Appold, a British engineer and inventor, introduced a curved-vane centrifugal pump. Finally, another British engineer, Osborne Reynolds (1842-1912), built the first turbine or centrifugal pump in 1875. Reynolds is more famous for his study of fluid dynamics, having the "Reynolds Number" named after him in relation to his studies on turbulence in water flow analysis.

In general, modern centrifugal water pumps operate at speeds much higher (e.g. 1800 or 3600 rpm) than were typically obtainable before the advent of steam or internal combustion engines and electrical motors. Therefore, centrifugal pumps were not technologically feasible or commercially viable before these devices were invented and readily available.

1.6. MODERN FIRE PUMPS

Initially, the first industrial firewater pumps were of the wheel and crank reciprocating model that were driven by mill machinery, powered by a waterwheel or windmill. This arrangement was not very practical, because if the mill waterwheel or windmill stopped, the fire pump would also stop. The English engineer, Thomas Savery (c. 1650-1715), patented the steam pump in 1698 after Denis Papin developed a first crude model in 1690. These first steam-driven pumps were initially applied to remove water from coal mines in England, but were later adapted to a wide variety of uses including as firewater pumps for municipal and industrial applications.

The first steam engine in America was imported from England in 1753. It was used to pump water from a copper mine in New Jersey. In 1795, the first practical steam engine was manufactured in America by Oliver Evans of Philadelphia, Pennsylvania. He later improved on it in 1799 with a high-pressure steam engine. It was particularly suited to the needs of the "colonial" industries of the time. Steam generation soon replaced or supplemented waterwheels or harnessed animals as an industrial power source.

Up until the late 1800s, almost all industrial firewater pumping systems were supplied with reciprocating steam-driven water pumps. The reciprocating steam engine dominated power generation for stationary and transportation services for more than a century, until the development of the steam turbine and the internal combustion engine. These engines were of heavy cast iron construction, and had a relatively low piston speed (600 to 1,200 ft/m) and low turning speeds (50 to 500 r/min), but were available with capacities of up to 18,642 kilowatts (25,000 hp).

With the development and provision of automatic fire sprinklers, requiring a more reliable water source, rotary pumps that were connected to the waterwheel of the mill were used. When steam supplies were provided at these locations, it replaced the water drive for the pumps and the reciprocating steam pump came in to use. As a result, the "Underwriters duplex", a double acting, direct steam-driven pump was universally provided as the standard fire pump for industry. As the name implies, these pumps were endorsed by the insurance carriers of the time and therefore were quite popular with industrial users.

When practical, large capacity, electrical motors and internal combustion engines became available in the early 1900s, the centrifugal pump came into full industrial use. Internal combustion engines or motors were readily applied as the driver of centrifugal firewater pumps due to their high speed of rotation and ease of installation.

Today, the centrifugal firewater pump is considered the most practical type of pump. It has the compactness, reliability, low maintenance, hydraulic characteristics and flexibility that have made earlier pump types obsolete for firewater use. Centrifugal firewater pumps are routinely specified for the protection of industrial facilities worldwide. They are found in both onshore and offshore facilities and may even be located underground.

1.7. MUNICIPAL WATER PUMPING PLANTS AND MAINS

Ancient civilizations generally used water buckets or large "syringes" to carry water from rivers or wells to a fire. When no readily available source was available, they probably did what firemen in London did in the early fourteenth century—they dug a hole in the street and waited for it to fill with ground water.

In 1562, the first municipal pumping waterworks was completed in London, England. A waterwheel pumped river water to a reservoir about 37 m (about 120 ft) above the level of the River Thames. Water was then distributed by gravity from the reservoir through lead pipes to buildings in the vicinity. By the late 1700s, steam engines pumped water in most European cities. The first water pumping plant to supply water for municipal purposes in the Americas was installed in Bethlehem, Pennsylvania in 1755. The water was pumped into a water tower through wooden pipes made from hemlock logs.

Wooden logs had been in use as a method for suppling water since the Middle Ages. Hollow bamboo was also used in the Far East (the Chinese even used bamboo pipe to transmit natural gas to light their capital, Peking, as early as 400 BC). After the decay of the Roman Empire, the Church took over the responsibility for supplying water and maintaining the old Roman aqueducts in some areas. Because of the tremendous size of the task, the aqueducts fell into disrepair after a few hundred years; however, under the protection of the Church, a guild of specialists in water supply had been created. Their technology spread over Europe alongside the simultaneous spread of the monastic orders. Their system of water pipelines from a source of supply, was made from hollowed-out logs connected with a cast iron collar, or the narrow ends of some logs fitted into the wider ends of others. Inside buildings, pipes of lead or bronze were used, a carryover from Roman days (lead is a poisonous material, and it is suspected that lead poisoning was a common cause of death in Rome).

The first municipal water supply system in America was built in Boston, Massachusetts in 1652. A series of wooden pipes was used to convey the water from a nearby spring to a central reservoir. By 1800, some 16 American cities had water-supply systems. Primitive fire hydrants on public mains in America began to be installed in the 1830s and 1840s. Prior to this time, there were some wooden water pipes with plugs at intervals which could be removed to obtain fire-fighting water.

There is evidence that cast iron pipes were used in a German Castle in 1455 and were also in use at Versailles, France around 1600. Cast iron pipes for city water mains in the United States were first used in 1817 in Philadelphia, Pennsylvania. The pipeline was 122 meters (400 ft) long and was 11.4 cm (4 $^1/_2$ inches) in diameter. The pipes were imported from England and were such an improvement on the existing wooden pipes (wooden pipes tend to rot and cannot hold much pressure) that the city decided to adopt them for all future installations. It has been stated that the limitations of wooden pipes hampered early attempts to pump water by steam and held back water supply technology in general.

In the eighteenth century, metal pipes were manually made and could withstand only a limited amount of pressure. The advent, in the nineteenth century, of steel pipes greatly increased the strength of pipes of all sizes. Initially, all steel pipes had to be threaded together. This was difficult to do for large pipes, and they were also apt to leak under high pressure. The application of welding to join pipes in the 1920s made it possible to construct leak proof, high-pressure, large-diameter pipelines. Since then, carbon steel pipe fire mains have been routinely provided, that were cement lined to limit internal corrosion. Unfortunately for some systems, the cement linings have been found to deteriorate after many years due to improper or poor initial application or materials, high water velocities or aging of the system. Copper-Nickel alloys (i.e. Kunifer 90/10) have found favor for use offshore since the 1970s due their high corrosion resistance and low weight.

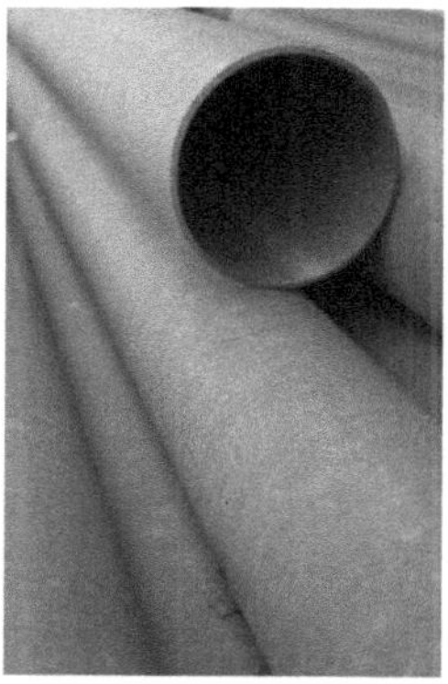

The latest trend for industrial facilities is to use reinforced fiberglass piping (e.g. Reinforced Thermosetting Resin (RTR)) for underground firewater mains and specialized fire-rated (including protection against jet fires) fiberglass materials for all the firewater piping on offshore structures. This offers the advantage of superior corrosion resistance and the weight saving desired for offshore facilities (fiberglass pipe weight is approximately one sixth that of steel piping). A very large wall thickness and water flowing through the pipe, allows the fiberglass pipe to withstand hydrocarbon fire exposures, including jet fires, for a limited duration.

Almost every city and town in America has been provided with municipal waterworks, most of them publicly owned and operated. These public water mains serve to provide water to industries and communities for domestic consumption and also for fire-fighting water. Additionally, hydrants are routinely provided on all city and private fire mains for the purposes of fire-fighting support.

1.8. OFFSHORE FACILITIES

Firewater pumps for offshore oil and gas installations need to be submerged in water because of the high level of the work platforms above the sea level and inadequate lift available for the pumps if located at the platform level (see Figure 1-2). Therefore, "down-hole" vertical turbine line shaft firewater pumps were commonly adopted as an extension of onshore well pumps during the beginning of the offshore industry and until the early 1980s. They are a standard type of pump used to supply offshore firewater systems. Several variations in drive and configuration are used to improve economics, ease of installation, weight impacts and reliability (See Chapter 9).

As offshore platforms moved into deeper waters and harsher environments, they became more complex and the use of obsolete tankers or dedicated ships as offshore processing facilities evolved. The constraints of space, weight and electrical area classification of the facility had to be considered more carefully. Electro-submersible and hydraulic-driven pumps gained considerable favor for offshore use because the power could be supplied from a dedicated generator or hydraulic power pack positioned in a convenient location and some of the topside drive hardware to the pump could be eliminated, allowing a space or weight savings. Only high integrity and durable pump driving systems (i.e. electro-submergible or hydraulic motors) should

Figure 1.2 Gulf of Mexico Shallow Water Platform

be selected for underwater use, otherwise continual repair and downtime of the firewater pumping system will occur.

An electro-submersible pump or hydraulic drive system still requires a dedicated topside power generation system, which increases its cost compared to a directly driven diesel engine line shaft pump. Because of this, the use of diesel-driven line shaft pumps has again been favored, especially where marginal returns are expected from some oil and gas production fields.

2 PHILOSOPHY OF PROTECTION

2.1. INTRODUCTION

Before the consideration of the installation of a firewater pump is undertaken, the need for it should be firmly established. This need should be established on the basis on the fire protection or risk philosophy promulgated for the facility by senior or executive management during its design.

In some cases, a firewater system is not absolutely necessary in the protection measures provided for a facility. Therefore, a firewater pump is not an absolute requirement for all industrial facilities. Furthermore, other mechanisms may provide adequate sources of firewater flow that make the provision of a firewater pump irrelevant. It should also be remembered that a fire suppression system is used in the last stages of a fire emergency or explosion event. Other highly effective fire protection measures may already have brought an incident under control (e.g. process emergency shutdown and isolation, depressurization, blowdown, etc.) before a fire suppression system is necessary.

Also, remote, non-critical or low value facilities may not require a firewater system since the cost of protecting these facilities outweighs their value to the organization.

2.2. PROTECTION OPTIONS

2.2.1. PROCESS EMERGENCY CONTROL MEASURES

The primary process emergency control methods are process controls, isolation and depressurization. These measures, if properly provided, should control and extinguish an incident relatively quickly. Although these systems provide for process control and incident minimization, they do not cater for fire control and suppression needs. Therefore, additional measures must be provided to accommodate the fire protection aspects required.

2.2.2. INCIDENT FUEL CONSUMPTION

One avenue of protection for a facility is to allow a fire incident to resolve itself by allowing the fire to consume all the available fuel to it. This allows conservation of water supply sources to those facilities that have not been damaged and require cooling water or exposure protection. It is the simplest and most economical of protection measures, however the incident itself may cause considerably more damage unless it is terminated at the earliest opportunity (i.e. additional product and facilities may be destroyed). In some instances, the amount of fuel available to an incident may preclude the consideration of this option (i.e. well blowout, large storage tank inventory, etc.), as

Fire Fighting Pumping Systems at Industrial Facilities. DOI: 10.1016/B978-1-4377-4471-2.00002-5

the fire may continue to burn for a considerable time, resulting in additional factors for consideration (i.e. pollution, public response, long term reputation, business interruption losses, increased cost of incident control, etc.).

2.2.3. PROVIDE PROTECTIVE MEASURES

The best option is to provide some amount of protection features to an installation based on the cost/benefit, or commensurate with the risk involved. National or local codes may also require the installation of fire protection devices. These are usually a combination of passive and active systems. This provides a balanced approach to protect very high risk areas with suitable protective measures, while lower risk areas receive less installed protective measures.

2.2.4. PASSIVE SYSTEMS

Passive fire protection features are normally preferred over active systems due to the inherent safety they provide without the need for additional intervention by manual means, or for detection and control systems that may malfunction or be impaired due to the incident. The primary passive measures include spacing and installed protective barriers, limitation of fuel sources and utilization of inherently less hazardous processes. Passive systems cannot always be provided to some equipment, because of other inspections or conditions imposed on the facility, e.g. provision of vessel fireproofing versus the need to conduct accurate metal thickness checks for corrosion.

2.2.5. ACTIVE SYSTEMS

Active systems are provided to automatically or manually detect and apply fire protection measures. These systems usually employ an extinguishing agent that is used at the time of the fire incident. Common systems include fire hydrants, monitors and hose reels for manual applications, and automatic water spray and deluge fire protection or water exposure cooling systems (see Figure 2.1). Automatic systems are arranged in combination with detection and control systems. Firewater pumping systems form part of the active fire protection system provided for an installation.

2.3. INSURANCE REQUIREMENTS

Industrial insurance underwriters have a vested self-interest in preventing losses at facilities they insure, even though most large industrial facilities may have a large deductible. Evidence of application of common industry practices to avoid major losses (from small incidents) will be investigated during evaluation of the property during the annual surveyor's inspections or initial assessment of the facilities prior to issuing a policy. The insurer's surveyors' assessment and recommendations (if any) to reduce losses at the facility will be reviewed during the insurance premium determination. If a risk is considered below the normal standards expected for the type of plant

Figure 2-1 Example of an Active Fire Protection Water Spray System

under examination, an insurer can insist on the implementation of recommendations as a condition of insurance cover.

Provision of an adequate firewater system usually rates highly in the risk assessment for any facility. Historically, municipal insurance grading schedules have placed approximately 35 percent of their allowable deficiency rating points on an adequate firewater supply and distribution system during grading of insurance levels for a community. This is the highest percentage for any fire protection feature that they examine under their rating schedule. Insurance agents generally request the particulars of the firewater system for a facility they underwrite.

The economic benefits of the provision of a facility firewater system can be easily demonstrated by any insurance underwriter. The provision of an adequate firewater supply and distribution system provides a lower insurance premium over the life of the facility compared to the insurance premium charged for a location without such provisions. In fact, it may even pay for itself. If you consider in addition, the negative reputation, prestige, legal and public indignation effects where an incident does occur and an adequate firewater supply system has not been provided, the justification for not providing it is negligible or non-existent.

2.4. INTERNAL COMPANY POLICIES AND STANDARDS

Most companies have their own internal management policies and engineering standards or guidelines for the design and construction of their facilities. These guidelines are based on the accumulated experience of the company personnel for the most economical and prudent standard their particular facilities should be built to. Rather than requiring extensive research for the most preferable design each time a facility is constructed, they provide a cost effective reference to construct a facility. In many cases these requirements exceed or expand on the provision of safety measures required by regulatory agencies or insurance requirements. Rather than being restrictive, they provide a base from which additions or deletions can be accomplished when adequately justified. Such documents should be endorsed by company management.

These documents may be extensive and define if and merely when a fire pump is required, the installation and arrangements, firewater supplies, durations and types of pumps and drivers allowed, or they may be relatively simple and merely require the installation to meet basic local fire codes (i.e. National Fire Protection Association (NFPA) 20, Loss Prevention Council (LPC) Rules, etc.).

3 FIREWATER FLOW REQUIREMENTS

3.1. INTRODUCTION

The purpose of a firewater pump is to supply water in the proper quantities and pressures to meet fire protection water application requirements. Unless these fire protection requirements have been analyzed and defined, the quantity and pressure requirements for a firewater pump cannot be accurately specified. This is the first step in determining the fire pump design for a plant.

Fire flow is a common term in the fire protection profession for the required firewater delivery rate for a particular occupancy. The term was originally derived by the insurance industry (Insurance Services Office (ISO)), indicating that classified water flows or fire flows according to a prescribed formula. These fire flow determinations were generally for municipal occupancies that ISO underwrote with insurance or was requested to evaluate. The formula for the required fire flow was based on the size of a building and its construction type, i.e. wood frame, noncombustible or fire-resistive construction, etc. ISO did not support the use of its fire flow calculation procedure for regulatory purposes and elected not to make available key supplemental information that was necessary to properly apply its method. Therefore, the fire flow criteria in the Uniform Fire Code (UFC) uses a modified version of the ISO fire flow formula for code enforcement purposes that includes factors related to the types of construction in the Uniform Building Code (UBC) and "sustained attach" fire suppression efforts for post-flashover fires.

Additionally, the estimation for the required fire flow for buildings for the American Fire Service was researched by Fire Chief Lloyd Layman in the late 1940s. His work was originally published by the National Fire Protection Association (NFPA) and concluded that compartment fires (i.e. those in buildings) could be generally extinguished during *incipient* stages (before flashover), using a rule of thumb of "3.785 liters per minute per 2.8 cubic meters (one gallon per minute per 100 cubic feet) of volume in a burning compartment". Unfortunately, this fire flow estimation did not consider building construction materials, occupancy differences, ventilation openings and exposures, because Layman's research did not consider these to be relevant, contrary to current understanding. Additionally, post-flashover fires require substantially greater volumes of water for control and suppression than pre-flashover fires do.

For large industrial facilities, firewater pumps are expected to support a number of different fire hazards affecting the facility. Firewater demands are normally calculated on the maximum rate of water that will be needed at any one time and applicable to a single fire incident. This is because it is unrealistic to assume that the total plant area will be involved in a fire at one time, and that it is also impractical to meet water

Fire Fighting Pumping Systems at Industrial Facilities. DOI: 10.1016/B978-1-4377-4471-2.00003-7

storage, firewater pumping and distribution requirements for a simultaneous incident through a large industrial facility. Traditionally, it has been expected that the area most remote from a firewater pump would produce the most taxing conditions for the pump and therefore it would be used to define the firewater flow and pressure requirements. With today's large industrial onshore complexes and multi-storied offshore installations, all the facility firewater demands have to be analyzed in order to establish the worst-case firewater flow requirements.

All industrial fires are not exactly alike. This is due to the fuels involved, weather, arrangement of the facilities involved, and the nature of the fuel configuration available for combustion, i.e. spill, gas, pressurized jet release, etc. Consequently, numerous methods have been proposed and investigated to determine the quantity of water that is enough for flow and duration requirements. Recommended application rates are also found in national fire codes (e.g. NFPA 13, 15 and 30) for fire control, suppression and exposure cooling. Recent trends in the industry have begun to use computer simulations of fire events with corresponding theoretical estimates of fire densities to control and suppress such incidents.

When materials are burned, they each have their individual heat releases. Water has the capacity to absorb large amounts of heat and convert it to steam which is also useful for fire smothering purposes. Therefore, the amount of water for control and extinguishing of certain fire exposures and also for extinguishment has been roughly estimated through the examination of this principle and through experimentation.

3.2. RISK AREAS

Common practice in the petroleum and chemical industry is to separate process areas into individual risk areas by means of physical separation of 15 meters (50 ft) or more or by the provisions of hydrocarbon rated fire barriers (see Figure 3-1). Tank farm risk areas are analyzed separately, generally according to NFPA 30 for tank fire control (or group of tanks) and adjacent storage tank cooling requirements.

The risk areas are then assigned firewater flow requirements based on the hazard they contain and fixed suppression systems. These risk area flow rates are based on common industry practices or the requirements found in fire codes for the control, suppression and cooling of chemical or petroleum fires and exposure protection from adjacent facilities. The risk area with the largest firewater demand will then become evident. Therefore, the capacity of the firewater pumping system should be based on the largest single process unit fire, defined offshore fire risk area or largest tank fire scenario.

Since fixed firewater monitors are the basic unit of protection for open industrial process areas, the first step is to determine the maximum number of firewater monitors that will be used in the risk area fire incident. This number times 1,892 l/min (500 gpm) for each monitor capacity will provide the fire flow for the monitors. Added to this should be the allowance for hand lines and for the provision of fixed water spray

Figure 3-1 Typical Complex Petrochemical Facility

systems. Similarly, for enclosed areas such as offshore modules which are defined as a single risk area, a density can be defined and a total flow rate estimated.

Example (onshore facility):

Five monitors at 1,892 l/min (500 gpm) = 9,462 l/min (2,500 gpm)

Four handlines at 946 l/min (250 gpm) = 3,785 l/min (1,000 gpm)

Deluge Sys. 1,892 l/min (500 gpm) = 1,892 l/min (500 gpm)

Total Firewater Flow 15,139 l/min (4,000 gpm).

3.3. EXPOSURE COOLING REQUIREMENTS

Some locations may be exposed to fire conditions from adjacent areas. These exposures, although separated to prevent flame carryover, may still allow radiant heat to be transmitted. The radiant heat effects may weaken or ignite nearby structures. Exposure cooling water is therefore required in these instances. This may be provided by manual means or fixed installations.

3.4. FIRE CONTROL REQUIREMENTS

Firewater may be needed to control the burning of a fire, rather than to extinguish it. An example of this would be for a gas leak, where the extinguishment of the flame would lead to a vapor cloud formation and subsequent explosion if ignited. In a case like this, the gas leak would be left to burn until the fuel supply was isolated and the surrounding exposures kept cool through cooling sprays.

3.5. SUPPRESSION REQUIREMENTS

All fires need to be suppressed by some means. The most common is through the application of water supplies. Suppression requirements can be accomplished through manual or automatic methods or a combination of both. The quantities of water have to be determined in advance through system hydraulic design analysis for fixed systems and pre-fire planning for manual methods. These amounts can be determined from recommended application rates found in local and national fire codes and company practices.

3.5.1. EGRESS WATER SPRAYS

Some high-hazard manned locations require an egress water spray protection. This spray cools the occupants and acts as a heat shield against radiated fire sources during their emergency evacuation. The author was involved with the design of an egress water spray for the access gantry provided to the Space Shuttle at Kennedy Space Center for the egress of the astronauts from the spacecraft in an emergency.

3.6. RESIDUAL PRESSURE REQUIREMENTS

During flow of the required quantities of water for fire-fighting, a residual pressure should be maintained in the most hydraulically remote or highly demanding system to ensure that an adequate density or reach of firewater streams is maintained for all application devices.

Additionally, most localities require that a residual pressure of 140 kPa (20 psi) be maintained in the city municipal water supply mains. This pressure is required because it is the minimum requirement for the supply of water to fire department pumper trucks, to prevent collapse of water mains or failure of fittings, and also to ensure that negative pressures do not occur on portions of the system that are at a higher elevation. Negative pressures will cause a backflow to occur in the system.

The hydraulic designs of most deluge and water spray systems require a specific residual pressure to meet the required performance. Additionally, most fixed firewater monitors, hose reels and hydrants for high hazard chemical or petroleum process areas need at least a 700 kPa (100 psi) residual pressure to provide effective coverage. NFPA Class I, II and III standpipes for buildings require a 455 kPa (65 psi) residual pressure with the required water flow on the *highest or most remote* outlet for the system.

4 DURATION OF FIREWATER SUPPLIES

4.1. INTRODUCTION

At each facility, the amount of water readily available to fight a fire will determine the period of time over which firewater supplies are available. There are generally two categories of water supply available for firefighting—unlimited supplies (e.g. oceans, rivers, lakes, etc.) and those that are considered limited (e.g. storage tanks, reservoirs, public mains, etc.). Dependent on the type of water supply available to the facility for firefighting, decisions have to be taken on whether primary or supplemental water storage supplies need to be provided. Normally, those locations with unlimited water supply sources do not need primary or supplemental water storage facilities for firefighting water, as by definition adequate and unlimited water supplies are to hand.

Locations that do not have an unlimited source of immediate firefighting water generally provide a mechanism to supply adequate amounts of water for firefighting in an emergency from an onsite storage location or public water mains with a refilling mechanism.

4.2. CAPABILITY OF PUBLIC WATER MAINS

Where the public water main has been tested for the required fire flow of the plant, and it does not drawn down water pressures below 140 kPa (20 psi), the public main may be considered to be an unlimited source of water. The source of the public main and its storage amounts, along with daily and seasonal fluctuations, need to be evaluated in this case. Where the public main provides adequate water supplies, the question of its availability (i.e. duration) may not be a factor. However, most industrial zones within a municipal area require considerable amounts of water and cannot meet the requirements of firewater flows for industrial plants. Firewater storage tanks are provided to supply the necessary water supplies to control and suppress the worst fire incident at the facility.

4.3. PRIMARY SUPPLIES

Water storage supplies at a facility can be categorized as primary and reserve. The need for a reserve supply is dependent on the size of the facility and the philosophy of protection adopted. These aspects are discussed in the next section entitled ''Reserve Supplies''.

Fire Fighting Pumping Systems at Industrial Facilities. DOI: 10.1016/B978-1-4377-4471-2.00004-9

The amounts needed for primary or reserve storage depends on the type of hazard present, the expected duration of the fire hazard that will develop and the protection measures selected. The Worst Case Creditable Event (WCCE) will normally dictate the maximum firewater flow rate that is required for a facility. Since a facility is normally separated into risk areas (i.e. tank farm, process unit, utilities area, etc.), each of these risk areas can be identified with required firewater flow rates to meet the company's or regulatory requirements for fire control and extinguishment. Some hazards have specific flow rates and durations specified by the National Fire Protection Association (NFPA) which can be followed (i.e. protection of storage tanks, automatic sprinklers, etc.). Once the flow rate has been determined, duration of water supplies can be applied that are recommended by insurance guidelines, industry practices, company polices or the expected duration of the fire. The NFPA Handbook (20th Edition, Table 15.2.6) recommends increasing durations for increasing amounts of fire flow rates. Table 4-1 provides recommended duration of fire flow based on the NFPA Handbook recommendations.

Table 4-2 shows a survey of some prescriptive firewater durations identified in NFPA fire codes. The survey shows that there is no common duration, durations vary with the type of hazard and they are written as a prescriptive requirement rather than as a risk-based approach.

Table 4-1 Firewater Duration*

Fire Flow Rate	Duration	Storage
3,785 – 9,463 l/min (1,000 – 2,500 gpm)	2 hours	1,192,801 L (300,000 gals.)
11,355 – 13,248 l/min (3,000 – 3,500 gpm)	3 hours	2,384,550 L (630,000 gals.)
15,140 – 17,033 l/min (4,000 – 4,500 gpm)	4 hours	4,087,800 L (1,080,000 gals.)
18,925 – 20,818 l/min (5,000 – 5,500 gpm)	5 hours	6,245.250 L (1,650,000 gals.)
22,710 l/min (6,000 gpm)	6 hours	8,175,600 L (2,160,000 gals.)
26,495 l/min (7,000 gpm)	7 hours	11,127,900 L (2,940,000 gals.)
30,280 l/min (8,000 gpm)	8 hours	14,534,400 L (3,840,000 gals.)
34,065 l/min (9,000 gpm)	9 hours	18,395,100 L (4,860,000 gals.)
37,850 - 45,420 l/min (10,000 to 12,000 gpm)	10 hours	27,252,000 L (7,200,000 gals.)

*NFPA Handbook, 20th Edition

Table 4-2 Survey of NFPA Fire Codes for Firewater Durations

NFPA Code	Reference	Duration
13 (2010)	Table 11.2.2.1	$^1/_2$ to 1 $^1/_2$ hours
13 (2010)	Table 11.2.3.1.2	$^1/_2$ to 2 hours
13 (2010)	Table 14.1.3	$^3/_4$ to 2 $^1/_2$ hours
13 (2010)	Table 15.1.1	2 to 2 $^1/_2$ hours
13 (2010)	Table 16.2.1.3.5	$^3/_4$ to 2 hours
13 (2010)	Table 16.2.2.1	1 $^1/_2$ to 2 hours
13 (2010)	Table 16.3.1.3	1 $^1/_2$ to 2 hours
13 (2010)	Table 17.2.1.3.1	2 to 2 $^1/_2$ hours
13 (2010)	Table 17.3.1.3	2 hours
13 (2010)	Table 17.3.2.1	1 $^1/_2$ hours
13 (2010)	Table 18.4 (d)	1 to 3 hours
13 (2010)	Para. 22.7.5	1 hour minimum
14 (2010)	Para. 9.1.4.1, 9.2, and 9.3	$^1/_2$ hour
15 (2007)	Para. 7.2.3.1.4	1 hour minimum
15 (2007)	Para. 7.4.3.8.4	1 hour
15 (2007)	Para. A-4-2.8	4 hours maximum
20 (2010)	Para. 4.6.4.1	Design duration
30B (2011)	Para. 4.6.4.1.1	2 hours
101 (2009)	Para. 32-2.3.5.3 (2)	Authority Having Jurisdiction
307 (2011)	Para. 7.2.3	4 hours
409 (2011)	Para. 7.85	$^1/_2$ hour
430 (2004)	Para. 4-11.4.3	2 hours
750 (2010)	Para. 10.3.1	$^1/_2$ hour minimum
801 (2008)	Para. 5.10.2 (3)	$^1/_2$ hour minimum
804 (2010)	Para. 11.7.3	2 hours minimum
850 (2010)	Para. 15.7.3	2 hours minimum
851 (2010)	Para. 6.2.2	2 hours minimum
851 (2010)	Para. 8.7.4	2 hours minimum
1141 (2008)	Para. 3.3.20	2 hours
1142 (2007)	Para. G.2	2 hours

If duration guidelines in the NFPA Fire Protection Handbook are used instead of the requirements cited in NFPA fire codes, i.e., say 10 hours for a fire flow of 45,420 l/min (12,000 gpm) instead of 4 hours, 27,252,000 litres (7,200,000 gallons or 171,000 bbls.) is required in storage instead of 10,900,800 (2,880,000 gallons or 69,000 bbls.), a 250 percent difference. Water storage tanks of these sizes are relatively unknown at industrial facilities which require their own self-contained water supplies, and can only be obtained if natural water supplies of sizable amounts are immediately available. Therefore, careful evaluation of the fire hazard and risk (for fire flow and duration) should be undertaken before firewater storage requirements are arbitrarily stated.

Table 4-3 Survey of Industry Duration Requirements

Reference	Duration
American Petroleum Institute RP 2510 (2001) Para. 10.3.1.5	4 hours
Factory Mutual Handbook	4 hours
Industrial Risk Insurers	4 hours

Fire duration and therefore water supplies can also be estimated from the amount of combustible material that is available in the process (see Table 4-3). If a facility has defined emergency isolation points for the control of the processes piping into and out of it, a reasonable estimation can be made for the amount of materials that would be contained within the process during an emergency. Since burning rates are relatively well-known for most materials, estimations can be made for various release scenarios (i.e. spills of different sizes, gas fires, etc.), and for the time taken to consume the material by fire considering no firewater intervention or with expected manual and automatic intervention. These periods will provide a rough approximation of the duration of the firewater required for each risk area.

4.4. RESERVE SUPPLIES

Reserve water supplies have sometimes been referred to within the industry. In general, this is a reserve capacity of water maintained for subsequent fire exposures in the same or another location of the facility that occur immediately after the initial or main water supply has been exhausted on a previous fire, and no other adequate sources of firewater are immediately available. In most cases, an inadequate source is one that cannot be replaced within 24 hours.

A reserve supply is basically needed because the mechanism to replenish the exhausted main supply does not have the capacity or sufficient time is not available to refill the main water storage tanks simultaneously during an initial incident without affecting the fire-fighting capabilities of other areas of the same facility (which have not shut down because of the incident). Therefore, it is more common for larger plants which have several operating areas that would not normally be affected by fire in another area of the facility (i.e. they have several large independent risk areas).

Also, this requirement may not be applicable to all facilities that have several independent risk areas, due to the philosophy of protection adopted for the facility.

It may be that when any large fire occurs at a facility, the policy will be to shut down the entire plant until adequate firewater supplies are available or regardless if firewater is available, or the entire plant may shutdown until the fire incident is considered completely extinguished or fully investigated, etc.

5 SOURCES OF FIREWATER PUMP SUPPLY

5.1. INTRODUCTION

The availability of local water supplies generally determines the source of supply for a firewater pump due to economics and reliability. Ideally, firewater pumps should always be provided with a water source under a positive pressure as this improves the reliability of the system and reduces wear. Certain locations and economic considerations, however, preclude the availability of positive pressure supply water to a firewater pump in all cases and lift pumps are used in these circumstances.

Since water is a resource, the removal or re-circulation of waste firewater from pump testing and training purposes for reclamation to the original source of supply should also be considered in determining the optimum and most economical supply source. Regardless of the source of water, it must meet the minimum requirements under adverse conditions.

Sources of water are generally classified as natural or manmade (constructed) sources. Natural sources are those contained and supplied by the ambient earth and include ponds, lakes, rivers, springs, artesian wells, seas and oceans. Constructed sources include aboveground tanks, elevated gravity tanks, wells, reservoirs and public water mains. Additionally, supplies can be classified as either groundwater or surface water. Surface water generally contains more bacteria, algae and suspended solids, while groundwater sources may contain significant amounts of dissolved minerals and entrained gases.

Water supplies to fire pumps that can be provided through strictly gravity systems are preferred over mechanical systems that must rely on pumps—systems which can be subject to failure.

5.2. SEAS AND OCEANS

The open sea or ocean can be considered an unlimited source of water, provided the pumping system is adequate. Although seawater is essentially free, it is aggressively corrosive to most exposed common metals. If a firewater pump is to be used for any extended time in saline conditions, corrosion-resistant materials must be selected for its construction. Most fixed, offshore structures that utilize seawater must usually lift it to 25 to 40 m (≈80 to 130 ft). Normal seawater conditions can vary from dead calm to wave conditions of −5 m to +10 m (16 to 33 ft) every 10-15 seconds. They also must maintain supply capability under extreme storm conditions (i.e. 100-year wave), when for some locations the seawater level may fluctuate from −10 to +20 m (≈33 to

Fire Fighting Pumping Systems at Industrial Facilities. DOI: 10.1016/B978-1-4377-4471-2.00005-0

66 ft.). Tidal fluctuations must also be accounted for at suction from open bodies of water, such as at offshore structures, pier locations and other shore-based installations.

Some facilities are provided with an area or a room that is "dry submerged" in the water itself. These are usually the submerged pontoons of semi-submersible vessels or the hulls of ships that have been permanently moored and are essentially a fixed installation. Seawater supplies to the fire pumps are provided through a seacock and therefore have a positive water feed to the pumps.

5.3. RIVERS, CHANNELS, PONDS AND LAKES

Natural occurrences of water supplies or even manmade enclosures provide a convenient source of water supply when these are located near the suction of the firewater pumping site. Climatic and seasonal changes must be taken into account when considering the reliability of the supply. Periodic or unexpected drought periods may make some prospective sources unfavorable. For supply from a river or stream, the quantity to be considered available is the minimum rate of flow during a drought with an average 50-year frequency. Calm and deep supplies are preferred over shallow and turbulent sources of water. Silting up or shifting of rivers and channels may also affect their adequacy. Suction screen or approved strainers are provided with impoundment crib for the pump suction location. The use of either traveling or double removal screens is preferred.

5.4. WATER WELLS (NATURAL UNDERGROUND RESERVOIRS)

Direct suction of water from underground water reservoirs or aquifers for use in a firewater pumping system is a convenient source of water, if the supplied flow rate can support the fire risk demands. Water wells with inadequate capacities for the defined risk may be used to supply storage facilities, which can then be used to supply firewater systems through firewater pumps or from gravity flow from storage tanks.

The capacity of a well water source is dependent on the aquifer permeability; its geology to sources of replenishment, such as rainfall and surface runoff, and other users of the aquifer itself. Agricultural users tend to be the main consumers of underground sources of water in rural areas. An aquifer performance analysis may be performed to determine the amount of water available from a given field and the necessary well spacing to avoid interference with other users.

Underground sources of water can contain many different types of contaminants which may cause corrosion, and may produce harmful or dangerous conditions for personnel. These include, but are not limited to, pH level, salts, such as chlorides, and entrained gases such as hydrocarbon gases, Carbon Dioxide (CO_2) and Hydrogen Sulfide (H_2S). Where the quality of water is unknown or questionable, it should be analyzed before a firm commitment is made to use it. Where there is a possibility of clogging of the well, necessitating cleaning and overhauling the well pump, periods of unavailability must be considered.

Dependence on a single water well, even when historical records indicate a favorable history of water supply availability, may be considered a feature of unreliability since it may not be physically possible to confirm the actual condition (i.e. geological stability) of the aquifer. Additionally, National Fire Protection Association (NFPA) 20 recommends that fire pumps should not be installed in a well where the pumping water level exceeds 61 m (200 ft) from the surface of the ground when operating at 150 percent of rated capacity of the firewater pump.

5.5. MANMADE RESERVOIRS (IMPOUNDED SUPPLIES)

Manmade reservoirs range from large city dams to specially constructed, smaller, embankment-supported and plastic-lined surface reservoirs or fabric tanks.

Impounded supplies such as dams or runoff reservoirs are usually dependent on natural rainfall or surface runoff for replenishment (see Figure 5-1). Therefore, most very large reservoirs have lower water levels in late summer or, due to limited natural rainfall, may go through extended periods of low water levels. The quantity of water considered available for these sources is taken as the minimum available during a drought with an average 50-year frequency. Smaller reservoirs are typically filled from water wells. Exposed reservoirs must consider the effects of evaporation and collection of sediments from airborne dust and sand when they are located in arid conditions.

Manmade surface reservoirs may sometimes offer a more economical solution than of the construction of a ground level or elevated steel storage tank. Generally, surface reservoirs can be easily and quickly constructed with earth-moving equipment and lined with semi-skilled labor in contrast to the need for specialized engineering

Figure 5-1 Typical Dam Water Reservoir

designs and qualified welders for the erection of steel tanks. Some "fast-tracked" industrial projects have employed the use of a man-made embankment reservoir as a temporary water supply to allow early plant startup, while the permanent water storage tank was still being constructed.

5.6. STORAGE TANKS

Water storage tanks can be either elevated, at ground level or buried. Most water storage tanks are built at ground level on hilltops higher than the service area. In areas with flat topography, the tanks may be elevated above ground on towers in order to provide adequate water pressures, or ground-level storage tanks with booster pumping may be provided. Marine vessels also used pressurized water storage tanks to supply firewater systems. Elevated and ground tanks have a preferred advantage over buried tanks due to the capability of these tanks to supply water by gravity (head pressure) to the firewater pump or even bypass it if the need arises (see Figure 5-2). Elevated and buried tanks are generally more expensive to provide than ground level tanks. Frequent cleaning of storage tanks may be a factor affecting reliability.

Storage tanks should have a means of automatic refill once a preset low level has been reached. A refill time of no more than 24 hours is recommended. Low level

Figure 5-2 Typical Elevated Water Storage Tank

alarms to protect the pumping unit should be provided. All tanks should be provided with measures to prevent freezing. Separate discharge points are recommended when a water tank supplies both domestic, operational and firewater needs. Domestic and operational water takeoff points are located above the maximum firewater supply requirement provided in the tank in order to prevent them from drawing down the reserve supply firewater storage in the tank.

Pressure tanks are used for firewater supplies in limited, private, fire protection services or in ships. The arrangement of firewater systems in ocean-going ships are normally covered by the conventions of the International Maritime Organization's (IMO) Safety of Life at Sea Regulations (SOLAS), rather than as a fixed installation covered by National Authority safety standards.

Vessels of 500 gross tonnage or more are to be provided with an international shore connection for hook-up to an onshore firewater supply system when docked. Onshore firewater systems that are provided with an international shore connection must be capable of handling the largest vessel that is expected to dock at the facility.

5.7. MUNICIPAL AND PRIVATE FIREWATER DISTRIBUTION MAINS

The most convenient source of firewater, when it is available, is the municipal water supply system. Where firewater is provided by a municipal water system, it may require the installation of an approved backflow prevention device (i.e. double check valves or break tanks) to prevent contamination or health hazard to the public water supply. Public water supply mains with inadequate water for the defined risk may be used to supply storage facilities, which can then be used to supply firewater systems through firewater pumps or gravity flow.

When considering the use of public water mains as a source to supply firewater pumps, the following possible variances should be reviewed in order to determine the relevant impact:

1. **Hourly Variances**: Water supplies from public water mains can vary from hour to hour due to their own supply and pumping arrangements. The maximum hourly demand is the rate of use during the hour of peak demand on the day of maximum demand. This demand rate normally establishes the highest rate for design of the municipal system at peak municipal pumping periods. The peak demands in residential areas usually occur in the morning as well as early evening hours (just before and after the normal workday). Water demands in commercial and industrial districts, however, are usually uniform during a working day.
2. **Variances in Local Uses**: Public water supplies can be easily affected by local use, which varies from day to day and from season to season. The maximum daily demand is the total amount of water used on the day of the heaviest consumption of the year.
3. **Future Growth**: In rapidly growing building areas, demand on the water system will change dramatically in a few years, unless the municipal water system has kept up with provision of adequate supplies and distribution.
4. **Municipal Supply Changes**: There may be changes made to the municipal supply that affects its rated output in the future. These can be either positive or negative in nature.

5. **True Static Pressure**: The test static pressure for a municipal system may not be a true static pressure as it is really a residual pressure with some unknown residual flow from a multitude of municipal users. However, since a true static pressure in a system in operation may never be realized, it may not be significant to account for the unknown flows and assume the static pressure for the sake of flow analysis is that which has been recorded and averaged over a period of time.

Additionally, firewater pumps taking suction from a public water main should not reduce the public main pressure below 140 kPa (20 psi) when operating at 150 percent of rated capacity. This is to allow adequate suction pressures for mobile fire pumpers and to prevent the collapse of the buried firewater mains from the weight of overburden; which is an area of concern regarding the latest trend to use reinforced fiberglass materials for buried firewater piping.

High firewater demands for high-hazard industrial locations cannot usually be met by the municipal water supply system, or a municipal firewater system may not be available or meet reliability requirements. Therefore, these facilities are provided with their own independent firewater storage systems supplemented by the municipal system when it is available.

5.8. SPECIALIZED OFFSHORE RAW SEAWATER SYSTEMS

Some offshore petroleum industry mobile exploration and production vessels have unique facility hazards that complement the use of raw seawater in the protection of the facility. On jack-up rigs and some semi-submersible vessels (also referred to as Mobile Offshore Drilling Units (MODU)), the firewater pump suction is typically fed from a "raw seawater system".

The raw seawater system starts from a pump in a submerged pontoon in a semi-submersible drilling platform, or from a water tower in a jack-up rig (see Figure 5-3). The main use of raw seawater is buoyancy control and for well-drilling operations. Due to the nature and criticality of raw seawater systems in these facilities, they are

Figure 5-3 Semi-Submersible Drilling Platform

at least the same, or more critical to the safety of the facility (i.e. protection against blowout and loss of buoyancy or stability) as a firewater pump. The fire main is normally kept at the supplied pressure of the raw seawater system. When required, the rig firewater pumps boost the pressure of the firewater system from the raw seawater system. The arrangement of firewater pumps on these floating installations (e.g. vessels) are normally covered by the conventions of the IMO's SOLAS Regulations, rather than being considered as a fixed offshore installation covered by National Authority safety standards.

5.9. FIREWATER USAGE BY OTHER SERVICES

Non-fire protection related connections to the designated and reserved firewater pumps, supplies and distribution system should be avoided. Process, utility, domestic and other uses during a fire situation from the firewater supplies may unexpectedly draw down the specified firewater quantities so that a major fire incident cannot be controlled or suppressed. Using firewater for process operations or utility services may also lead to contamination of the firewater, or cause a process upset when large quantities of firewater are used and it cannot adequately feed the tapped-in process users. NFPA 24 and most internal company standards recommend that no connections be made to firewater hydrants other than for firewater use. Instances have occurred where the firewater system was contaminated with hydrocarbons because of cross connections to a process system and the firewater system inadvertently applied both water and fuel simultaneously to a fire incident.

As built drawings should be prepared and maintained for the firewater system. Proposals to connect other users to the system should be critically examined and alternative measures evaluated before routine allowance to allow the use of firewater for other services or process is considered. Generally, considerations such as cost for alternative sources, proposed flow rate, shutdown during an emergency, back-feed potential and other features are presented during an evaluation of the proposed connection.

In some specialized cases, firewater pumps use a backup feed system for *emergency* process cooling requirements, but not as the primary supply. Suitable controls are provided on the connection to ensure a prompt shutdown during a fire emergency condition. If the firewater system is allowed to be periodically or routinely used for purposes other than fire fighting (although this is not preferred), the pressure maintenance (i.e. jockey) pump should have sufficient capacity for such use, or a separate service pump should be provided. The service pump would not need to meet the requirements of a fire rated pump. In one aspect, this arrangement is beneficial as it provides confirmation that the firewater distribution system is available, flushes the system and provides another pump as a backup unit. However, the concerns mentioned previously regarding impairments to the firewater system by non-firewater consumers are still valid.

Combined firewater and utility water systems are sometimes found, where it is uneconomical to provide separate firewater and utility water distribution networks throughout the facility. At the time of an incident, all non-critical water users are shutdown. This type of system has to account for lower water pressures in the system during a fire incident and the impact this will have on critical users of the system. The method of shutdown for non-critical users has to ensure these will be isolated in a timely and effective manner.

5.10. EMERGENCY WATER SOURCES

Other available and recognized water sources which should be considered if the primary and reserve supplies become unavailable, include open bodies of water such as water from rivers, jetties, utility water systems, cooling towers, drainage sewers (or sewage treatment ponds) and swimming pools. Most of these sources are nominated for use by mobile firewater pumpers or portable units rather than direct connection to fixed firewater pumping systems.

At some marine terminals or jetties, suitable connections are fitted to the firewater main to allow mobile or fire tugboats to supply sea water to the main in the event of a breakdown of the facility fixed fire pumps, or for use in boosting the pressure in the shore firewater main. These connections should be sized and arranged to provide the required flow of firewater for the highest risk area at the facility.

Some of these sources may be treated with chemicals or contaminated with hydrocarbons or other compounds. These materials can interfere with firefighting foam extinguishing agents. An examination of the suitability of potential emergency backup water supplies should be undertaken as part of pre-fire planning for a facility.

5.11. WATER QUALITY

The quality of the water used for firewater service should be examined to determine its potential for detrimental effects to the firewater pumping system. These effects may have a negative result on the operability and service life of a firewater pump or to the distribution system it supplies and the end devices (sprinklers, monitor nozzles, foam systems, etc.). The effects can stem from water hardness, corrosion, entrained gases, sediments or sludge, freezing conditions, marine crustaceans, algae or seaweed growth, pollutants, and submerged debris. Suitable precautions must be instituted where these conditions produce unfavorable conditions in the system.

Water has such a strong tendency to dissolve other substances, therefore it is rarely found in nature in a "pure" condition. Small amounts of Oxygen and Carbon Dioxide (CO_2) gas become dissolved in rainfall droplets. Raindrops will also collect tiny dust particles. As surface water drains, it picks up fine soil particles, microbes, organic material and soluble minerals from the ground. In lakes, bogs, and swamps, water may gain color, taste, and odor from the decaying vegetation and other natural organic matter. Groundwater will usually acquire more dissolved minerals than surface runoff, because of its longer direct contact with soil and rock. Underground sources of water can contain many different types of contaminants which may cause corrosion and might produce harmful or dangerous conditions for personnel. These include, but are not limited to, pH level, salts such as chlorides, and entrained gases, e.g. Methane, CO_2, and Hydrogen Sulfide (H_2S). In populated areas, the quality of surface water as well as groundwater is directly influenced by human activities and the effects of pollution.

Firewater systems do not require the use of potable water. Fresh water systems, however, are preferred over seawater because of corrosion control and there are therefore material costs implications associated with seawater. In many cases, where seawater or brackish water is used for a fire incident, the system is flushed and recharged with fresh water when it is available, to prevent or avoid corrosion in the complete firewater system. However, it is preferable to initially construct the firewater pumping and distribution system out of corrosion resistant materials.

Hard water supplies, i.e. those that contain large amounts of calcium carbonate, should not be a concern for a firewater pump provided it is regularly operated. Hard water supplies tend to cause the accumulation of mineral deposits in predominately stagnant water conditions. Firewater pumps which obtain their supply from hard water sources may encounter an accumulation of mineral deposits in small bore piping or cavities at low elevations, such as in pendant sprinklers, if their supply piping has been arranged such that they are fed from the bottom surface of supply header piping, rather than have a take-off from the top surface of the piping. These accumulations may cause these locations to plug over time. Sprinklers installed in such systems should be removed and checked yearly for possible accumulations. The term 'total dissolved solids' (TDS) is used to describe the level of dissolved particulates in water. Technically, it is a measure of the combined content of all inorganic and organic substances contained in a liquid in molecular, ionized or micro-granular (colloidal sol) suspended form. Total dissolved solids are only normally discussed in relation to freshwater systems, as salinity comprises some of the ions constituting the definition of TDS.

The TDS tolerable to the firewater pumping and distribution system in mg/L (ppm) should be compared to the geochemical composition of the water to be used. High TDS levels can cause scale build-up in pipes, valves, and filters, reducing performance and adding to system maintenance costs.

5.12. ENHANCEMENTS TO FIRE-FIGHTING WATER

A variety of chemicals may be added to firewater supply systems to improve its ability to control and extinguish a fire. These include wetting agents and foam concentrates. Wetting agents are more commonly used with manual or mobile fire-fighting efforts, while foam concentrate systems are used with both mobile and fixed firewater systems.

Wetting agents, which are added to water, can reduce its surface tension. Water with a wetting agent has more penetrating capability and facilitates the formation of small drops necessary for rapid heat absorption. Wetting agents are purely hydrocarbon surfactant mixtures that are proportioned at extremely low concentrations (i.e. one to two percent) to improve the wettability of the water being applied. Wetting agents are used primarily where a high degree of water penetration is required for a three dimensional burning mass such as a coal pile or a forest fire.

By adding foam-producing concentrates to water, a water foam solution can be produced to "blanket" a fire on the surface of a liquid and exclude oxygen from the fuel. Foam is used to extinguish fires in combustible liquids, such as petroleum, chemical and for fighting fires at airports, refineries and petroleum distribution facilities. Some chemical additives can expand the volume of foam up to 1,000 times. This high-expansion foam-water solution is useful in fighting fires in confined areas or difficult-to-reach areas because the fire can be smothered quickly with relatively little water damage. High expansion foam displaces the oxygen in an enclosure causing fire extinguishment. The most common firefighting foams for industrial locations tend to be low expansion foams for hazardous liquid fires. These have an expansion ratio of about one to 100. Pumps specified for the injection of foam concentrate additives require positive displacement capabilities because of the high viscosity characteristics of foam concentrates.

5.13. MARINE GROWTH

Internal marine growth in the firewater system components and intake points will hinder performance of the pump and may eventually lead to plugging of the system. Marine growth can consist of algae, aquatic weeds or marine invertebrates such as asian clams, zebra mussels and other small organisms that can fit into the system. Static water will allow growth of algae in sunlit conditions. Asiatic clams have spread to over 70 percent of the United States. Zebra mussels have been found in the Great Lakes area and its surrounding tributaries.

There are several methods to control marine growth in a firewater pumping system. These include use of screens, construction of materials to resist or prevent marine growth, protective coatings, electrical fields and biocide injection.

Suction screens are provided as the primary protection against the intrusion of marine invertebrates into a water supply system. Pump suction screens and even pump components constructed of copper or brass tend to promote less aquatic growth. Wherever screens are fitted, they will have to be periodically inspected and cleaned. Pump intake piping made of fiberglass materials prevents organisms from securing an attachment point. Protective coatings make it difficult for organisms to attach themselves to the surface of the equipment. Biocide and electrical fields can disturb the organism's capacity to sustain life. A concentration of 0.5 ppm of chlorine in fire protection reservoirs has been found to effectively treat mollusc larvae distributed by water movement.

5.13.1. BIOCIDE INJECTION

One common method to prevent marine growth within a firewater system supplied by ocean seawater with a vertical turbine pump is to inject biocide at the operating inlet of the pump suction bell. Figure 5-4 provides a typical example of a fitting provided at the pump suction for this purpose. The biocide may be either a proprietary supplied preparation or one generated at the facility. Typically, the biocide generator used in industry is based upon an electrolytic cell. Seawater is partially electrolyzed to produce a solution of chlorine, which then hydrolyses to hypochlorite.

The solution so produced is a powerful biocide. Before the use of biocide is seriously considered, regulations for the protection of the environment should be consulted to determine if such a system is allowable and if so, the allowable release rates of the chemical. The use of a biocide in doses necessary to eradicate crustaceans in closed bodies of water could be devastating to the entire ecosystem if this is not verified beforehand.

Distribution of the biocide is accomplished with a dosing pump and a network of small bore pipes. Due to the highly corrosive nature of the biocide, it is usually distributed in a titanium or plastic piping system. Since titanium is expensive, it is

Figure 5-4 Typical Sodium Hypochlorite Injection Ring Fitted at Pump Intake

limited to the distribution header at the pump inlet and plastic piping is provided to the remaining portion of the system (see Figure 5-4).

A sodium hypochlorite solution is extremely corrosive and must be stored in stainless steel or glass reinforced plastic (GRP) tanks prior to use. Due to the corrosive nature of this biocide, any leakage or accumulation of it on the external pipework should be carefully inspected and measures instituted to immediately prevent further release or accumulation. Cases have been recorded where a drip leak of hypochlorite onto the external surface of the fire pump discharge line over a period of time caused the piping to corrode. The pipework ultimately ruptured when the system was undergoing tests because of the unknown corrosion activity.

A check on the release rate of the biocide system should be made every six months to ensure neither too much nor too little is being released into the firewater system.

5.13.2. OTHER MARINE GROWTH CONTROL METHODS

Other methods to be considered in controlling marine growth in a fire water system include the following:

- Use of a water supply source that is not infected with marine organisms, such as well water, potable water or pretreated water.
- Implementing a water treatment program for the water supply that includes biocides, pH or a combination of both.
- Removal of oxygen from the water supply to control biological growth.
- Reliance on a tight system to deny oxygen and nutrients to support growth in the system.

5.14. FUTURE USE, SOURCES AND DEVELOPMENT

The water supply arrangements should be reviewed every five to 10 years to evaluate development and replacement needs in the future. A plan should be available that indicates the current and proposed utilization of the system, replacement needs and future additions or sources of water supplies.

6 PUMP TYPES AND APPLICATIONS

6.1. INTRODUCTION

Four general classes of pumps are available for liquids—reciprocating, centrifugal, jet types and other methods. Reciprocating and centrifugal pumps are the almost exclusively used for fire protection pumping systems and are discussed in this book. Jet pumps and other methods of pumping have not found application within fixed fire protection systems and are not a concern here.

There are two major types of pumps—dynamic and positive displacement. Dynamic pumps maintain a steady flow of the fluid. Positive displacement pumps contain individual portions of fluid that are enclosed before they are moved along. Both of these types are used in fire protection pumping systems. Dynamic pumps are used for water supply and positive displacement pumps are used for foam.

Centrifugal pumps, which have a relatively flat characteristic performance curve, are generally selected for firewater pumping systems. They can provide a steady, non-pulsating flow of water at a uniform pressure over a wide range of flows. They can also idle against closed valves for a certain period of time without damage to the pump or connected equipment. In some cases relief valves or governors are provided to prevent overpressuring the system.

Centrifugal pumps for firewater service can be driven by diesel engines, electric motors and steam turbines in various configurations. They are fed under positive pressure from aboveground supply sources or are submerged in the water supply such as in a well or ocean. Individually, applications within industrial facilities can range in size from 95 l/min (25 gpm) to as much as 47,332 l/min (12,500 gpm). They are considered a prime component in the protection afforded to an industrial installation and require specific examination for fire protection service suitability.

6.2. DYNAMIC PUMPS

6.2.1. CENTRIFUGAL PUMPS

Also known as rotary pumps, these have a rotating impeller, or vanes, which are immersed in the fluid being pumped, and are encased in housing to direct the water flow. The rotation induces an increase in pressure of the liquid by developing a centrifugal force (i.e., it imparts a mechanical force to the liquid). In fact, the centrifugal pump receives its name from its dependence on the centrifugal action of the rotating impeller for the discharge of a liquid. The impeller is a rotating disk that is provided with an enclosing circular casing. The fluid enters the pump near the center of the impeller, which sweeps it away with a circular motion to an outlet on the periphery of the casing.

As water enters the suction inlet of the pump and the eye of the impeller, the vanes of the rotating impeller pick it up. It is then discharged due to centrifugal force. A

Fire Fighting Pumping Systems at Industrial Facilities. DOI: 10.1016/B978-1-4377-4471-2.00006-2

vacuum is created at the eye of the impeller, due to the discharge of the fluid from the outlet of the pump casing. This vacuum causes additional water to flow into the eye of the impeller replacing the water that has been just discharged. The action is continuous for the period of pump operation provided that the following conditions are met:

1. No air enters the suction of the inlet of the pump.
2. The supply of water is adequate to supply the required discharge volumes.
3. The height of lift (or suction) is not too great for the volume of water being discharged.
4. The suction or supply mains is not too small to supply the desired volume.
5. There is no obstruction or restriction to limit the flow in the suction piping of the pump.

The impeller also gives the liquid a relatively high velocity that can be converted into pressure in a stationary part of the pump, known as the diffuser. In high-pressure pumps, a number of impellers may be used in series, and the diffusers provided after the impellers may contain guide vanes to reduce the liquid velocity gradually. For lower pressure pumps, the diffuser is generally a spiral passage, known as a volute, with its cross-sectional area increasing gradually to reduce the velocity efficiently. A liquid must surround the impeller when it is started, i.e. it must be primed. This can be accomplished by providing a check valve in the suction line. The check valve will hold liquid in the pump when the impeller is not rotating. If the check valve is not reliable, the pump may need to be primed by the introduction of fluid from an outside source. A centrifugal pump normally has a valve located in the discharge to control the flow and pressure.

For low flows and high pressures, the action of the impeller is largely radial as the centrifugal action governs the design. For higher flows and lower discharge pressures, the direction of the flow within the pump is more nearly parallel to the axis of the shaft and the pump is said to have an axial flow. The impeller in this case acts in a similar way to a propeller. The transition from one set of flow conditions to the other is gradual, and for intermediate conditions, the device is called a mixed-flow pump.

Most centrifugal fire pumps convert more than 70 percent of the velocity energy to pressure energy, which added to the pressure energy as discharged by the impeller, gives a high percentage of recovery.

Centrifugal firewater pumps include single and multistage units of horizontal and vertical shaft design. The nomenclature of pumps refers to the position of the impeller shaft in their description, for example a vertical turbine pump may be driven by a right angle gear drive to a horizontal diesel engine or an electric motor positioned vertically on the pump shaft. Since the pump is in a vertical position, whatever position the driver is in does not apply to the description of the pump. Approved or listed firewater pumps have capacities of 95 to 18,925 l/min (25 to 5,000 gpm), with a net pressure range from 280 kPa to 2,800 kPa (40 psi to 400 psi). Some industrial applications have used pumps with capacities up to 47,332 l/min (12,500 gpm) for firewater purposes.

The following is a brief description of the various types of firewater pumps that are in use.

- **Horizontal Split Case**: This pump type is characterized by a suction impeller with an inboard and outboard bearing with a horizontal orientation for the drive shaft of the impeller (see Figure 6-3). It is normally used where a positive suction is provided. It can be arranged to be mounted with the shaft vertically. The split case for a horizontal type has the advantage that the rotating element can be removed without disturbing the suction and discharge piping. An innovative design incorporates a "tilted parting" arrangement of the split case. This concept minimizes turbulence at the eye of the impeller by utilizing a straight laminar flow of water to it. This improves efficiency and also aids in saving pump space (see Figure 6-1A).

- **End Suction and Vertical Inline**: This pump type has either a horizontal or a vertical shaft with a single suction impeller and a single bearing at the drive end. They are useful for adapting to existing systems because of their orientation versatility and minimal installation requirements. Small vertical inline pumps can be supported with ordinary pipe supports on either side of the pump, eliminating costly foundations or pads (see Figure 6-1B).

Figure 6-1A Horizontal Shaft Fire Pump

Figure 6-1B Vertical Shaft Fire Pump
Source: Photo courtesy of Patterson Pump Ireland Limited. Reprinted with permission.

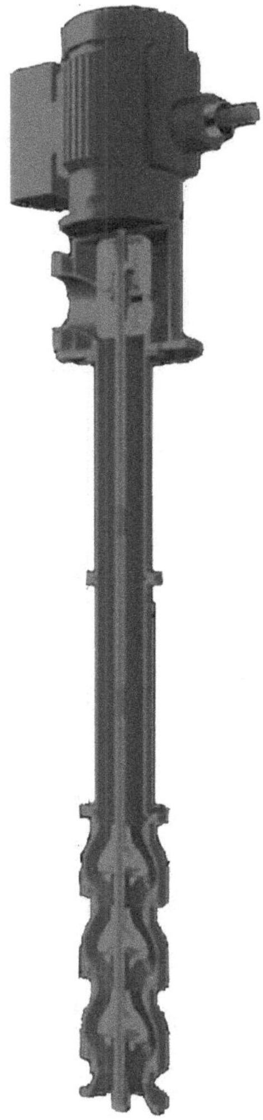

Figure 6-2 Vertical Shaft Fire Pump Types
Source: Photo courtesy of Patterson Pump Ireland Limited. Reprinted with permission.

- **Vertical Shaft, Turbine Type**: This pump has multiple impellers and is entirely submerged in the supply source. It is suspended from a discharge head by sections of column pipe. The column pipe also supports and guides the pump vertical drive shaft and bearings where it is driven from an electrical motor or diesel engine. This type is commonly used where a suction lift is required such as from a well, offshore caisson or water pit (see Figure 6-2).

6.2.2. PUMP AND IMPELLER DESIGN RELATIONSHIPS

The head produced by a centrifugal pump is a function of the square of the water velocity at the periphery of the impeller. It follows that the pressure achieved by a pump is proportional to the square of the speed of rotation and to the square of the

Figure 6-3 Interior View of Centrifugal Fire Pump
Source: Photo courtesy of Patterson Pump Ireland Limited. Reprinted with permission.

impeller diameter. This fact places practical limitations on the head that can be developed by a single impeller.

Pumps for higher head than can be achieved with a single stage design are built with several impellers in series, i.e., although the impellers are mounted on the same shaft, water passes from the first into the second, etc. Each impeller provides an incremental increase in pressure. Extremely high heads are therefore possible in centrifugal pumps. As most firewater systems operate at relatively low pressures, e.g. 1,050 kPa (150 psi), a single impeller is normally adequate. Multiple impellers are required where higher pressures are demanded from the firewater system, because of operational requirements or because of the high lift required from the water source (i.e. vertical turbine pumps in wells or offshore service usually have more than one impeller).

Standard lines of pumps are constructed to fit impellers of different diameters into a pump casing of standard design. Different capacities can therefore be easily obtained by changing the impeller size or the speed of rotation to suit the specific client's needs. By producing standard designs of casings, and fitting impellers of different designs and sizes, a manufacturer can reduce the overall costs of production over a wide range of pump sizes.

Certain physical relationships exist that allow the performance of a centrifugal fire pump to be predicated for a speed other than that for which the pump characteristic is specifically known. Certain physical relationships also exist that allow prediction of the performance of a pump if the impeller is reduced in diameter (within a limit dependent on impeller design) from the characteristics obtained at the larger diameter.

For a pump using a given impeller, but having a variable speed driver, the head is proportional to the square of the speed, the capacity is proportional to the speed, and the power is proportional to the cube of the speed.

These expressions can be stated in the form of an equation, as follows:

$$Q = Q_1 \times (S/S_1);\ H = H_1 \times (S/S_1)^2;\ P = P_1 \times (S/S_1)^2$$

S = New speed desired, in revolutions per minute.
S_1 = A speed in revolutions per minute (RPM) at which the characteristics are known.
Q = Capacity, in gallons per minute at desired speed S.
Q_1 = A capacity at speed S_1.
H = Head, in feet, at desired speed S for capacity Q.
H_1 = Head, in feet, at capacity Q_1 at speed S_1.
P = Brake horsepower, at desired speed S at H and Q.
P_1 = Brake horsepower, at speed S_1 at H_1 and Q_1.

At constant speed, the head is proportional to the square of the impeller diameter, the capacity is proportional to the impeller diameter and the power required is proportional to the cube of the impeller diameter.

These expressions can also be stated in the form of an equation, as follows:

$$Q = Q_1 \times (D/D_1);\ H = H_1 \times (D/D_1)^2;\ P = P_1 \times (D/D_1)^3$$

D = Cut-down impeller diameter, in inches.
D_1 = Original impeller diameter, in inches.
Q = Corresponding capacity with D impeller.
Q_1 = Capacity with D_1 impeller.
H = Corresponding head with D impeller at Q.
H_1 = Head with D_1 impeller at Q_1.
P = Brake horsepower with D_1 impeller at Q and H.
P_1 = Brake horsepower with D_1 impeller at Q_1 and H_1.

6.2.3. SINGLE AND MULTI-STAGE ARRANGEMENTS

Centrifugal pumps can be specified as single or multiple stage. In multi-stage pumps, two or more impellers are arranged to operate in series, i.e., from one impeller directly to another. The advantage of multi-stage pumps is that although the quantity of water is the same as for a single stage pump, the total head developed is a product of the head produced from one stage times the number of stages or impellers. Two or more stage pumps should be considered instead of the provision of two or more pumps in series as there is an economical advantage and also a saving in equipment space and maintenance requirements.

6.2.4. VOLUTE AND TURBINE PUMP CLASSIFICATION

There are two general classifications of centrifugal pumps—volute and diffuser or turbine pumps. In volute pumps, the impeller is surrounded by a spiral case whose outer boundary is surrounded by a smooth curve named a volute. In turbine pumps, the impeller is surrounded by diffuser vanes that provide gradually enlarging passages.

Because of its resemblance to a reaction turbine engine, it is called a turbine pump. Pumps of this type are usually provided with variable speed drivers or drivers with fluid couplings. Usually, single stage pumps are of the volute type and high-pressure pumps frequently employ diffusers or guide vanes.

6.2.5. AXIAL FLOW PUMPS

Axial flow pumps have a motor-driven rotor that directs flow along a path parallel to the axis of the pump. The fluid thus travels in a relatively straight direction, from the inlet pipe through the pump to the outlet pipe. Axial flow pumps are most often used as compressors in turbo-jet engines. Centrifugal pumps are also used for this purpose, but axial flow pumps are more efficient. Axial flow compressors consist of alternating rows of rotors and stationary blades. The blades and rotors produce an increase in the air pressure as it moves through the axial flow compressor. Air then leaves the compressor under high pressure.

6.3. POSITIVE DISPLACEMENT PUMPS

There are a variety of positive displacement pumps available to support fire protection needs. Primarily they are used to supply additives to water for the enhancement of fire-fighting efforts (i.e. foaming agents). They generally consist of a rotating member with a number of lobes that move in a close-fitting casing. The liquid is trapped in the spaces between the lobes and then discharged into a region of higher pressure. A common device of this type is the gear pump, which consists of a pair of meshing gears. The lobes in this case are the gear teeth.

In most of these positive displacement pumps, the liquid is discharged in a series of pulses and not continuously, so care must be taken to avoid resonant conditions in the discharge lines that could damage or destroy the installation. For reciprocating pumps, air chambers are frequently placed in the discharge line to reduce the magnitude of these pulsations and to make the flow uniform.

6.3.1. ROTARY PUMPS

There are three types of rotary pumps for liquid applications—gear pumps, lobe pumps and sliding vane pumps.

Rotary pumps are positive displacement pumps that are usually applied for pumping fluids that are highly viscous. In fire protection applications they are commonly employed for the insertion of foam concentrates into water systems in exact quantities for the protection of flammable or combustible fluids.

6.3.1.1. GEAR PUMPS

These pumps consist of two gears that rotate against the walls of a circular housing. The inlet and outlet ports are on opposite sides. Fluid is drawn into the clearance space between the meshing gears and its pressure is subsequently increased by the rotation of the gears acting together.

6.3.1.2. LOBE PUMPS

Lobe pumps resemble gear pumps but have two to four lobes (rounded projecting parts) in place of gears. They deliver a steady flow without pulsation.

6.3.2. SLIDING VANE PUMPS

The vane pump consists of a cylindrical casing with a small, internal rotor positioned off center. Spring-loaded vanes project from the rotor to the casing. As the rotor revolves, the volume of fluid enclosed by successive vanes is decreased, thereby increasing the fluid pressure. Such units can handle small amounts of liquid or gas. For high-speed pumps, the springs are unnecessary because the vanes are forced against the casing by the centrifugal force of the rotating shaft.

6.3.3. RECIPROCATING PUMPS

These consist of a piston moving back and forth in a cylinder with appropriate valves to regulate the flow of liquid into and out of the cylinder. These pumps may be single or double acting. In the single acting pump, the pumping action may take place on only one side of the piston, as in the case of the common lift pump, in which the piston is moved up and down by hand. In the double acting pump, the pumping action may take place on both sides of the piston, as in the electricity or steam-driven boiler feed pump, in which water is supplied to a steam boiler under high pressure. These pumps can be single stage or multi-staged, that is, they may have one or more cylinders in series.

6.4. FIREWATER PUMP CHARACTERISTICS

Preferred pumps for firewater use are of horizontal centrifugal type having a flat pumping characteristic (head versus capacity curve) in order to maximize the capacity output available from the pump. The minimum residual pressure at the extremities of the distribution system may set the discharge pressure with additional allowances for piping friction losses.

Fire pumps mainly differ from other commercially available industrial pumps (e.g. American Petroleum Institute (API) Standard 610) in having a relatively flat characteristic head curve. Control of fire incidents generally demands varied or large amounts of water, sometimes at several locations, at a relatively constant pressure level. Additionally, a flat pump curve allows parallel pump operations to occur more easily.

6.4.1. CHARACTERISTIC FIREWATER PUMP "CURVE"

Each firewater pump is rated for a specific capacity and pressure, e.g. 3,785 l/min at 1,050 kPa (1,000 gpm at 150 psi), which equates to its 100 percent duty point. The

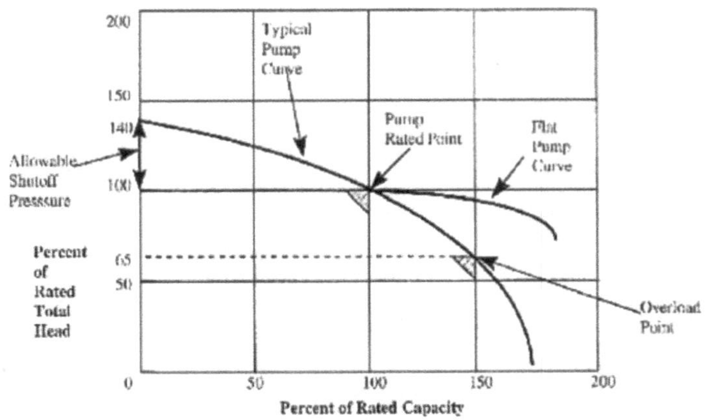

Figure 6-4 Characteristic Firewater Pump Curve

standard fire code requirement (i.e. National Fire Protection Association (NFPA) 20) also states that centrifugal firewater pumps are to have a stable characteristic curve and furnish no less than 150 percent of the rated capacity at no less than 65 percent total rated pressure. The total shutoff pressure or head for fire pumps is not to exceed 140 percent of the rated pressure (previously the maximum head for horizontal pumps was limited to 120 percent of rated head). The minimum shutoff pressure should also be no less than 100 percent of the rated pressure output of the pump at rated capacity (see Figure 6-4).

The main reason for these points is that they provide a relatively flat pump curve, so that the required pressure may be available across a wide range of water quantity demanded and that the pressure supply to a system is never less than what the pump is rated for when operating at less than rated capacity. In fact, tangents to a fire pump curve have a slope of zero or a negative value. Pressure availability is the critical feature in these pumps, for without adequate pressure, the water will not be induced to move in the firewater distribution system.

6.5. MAIN AND STANDBY FIREWATER PUMPS

Although NFPA 13, *Installation of Sprinkler Systems*, considers a single fire pump for a sprinkler system to be acceptable, a main and spare firewater pump should be provided for all industrial facilities, as the risk to a facility of not having an operational fire pump is normally too high to accept. The standby, spare or reserve firewater pump is available if the main or primary firewater pump is removed for maintenance, fails and even if there is an *over-demand* on the primary pumps. The main firewater pump is normally electrically driven, and the spare or backup firewater pump is usually provided with a diesel-driven driver for increased stand-alone reliability and availability as a backup unit.

The main pump(s) is the one selected to support major and catastrophic fire emergencies and is routinely started when the pressure in the firemain drops below the level which can be supported by the jockey pump or from other confirmed requirements. The main and backup pump sets are each sized for 100 percent of the required firewater flow. The capacity of the backup pump(s) should always be able to meet the requirements of the highest risk area similar to the main pump(s). In fact, the backup pump should always be thought of as a main or primary pump unit, though it may be configured to operate in a reserve or backup capacity.

In some locations, the primary and standby pumps are rotated in service, each one being the main and standby for specific periods to allow each to wear equally over the lifetime of the units. In these cases, a selector switch is provided to select which pump(s) will be the primary and which one(s) will be the backup.

In very critical locations and hostile offshore environments, such as the North Sea, three separately located 100 percent capacity firewater pumping systems are normally required or demanded by safety regulations, one main pumping system and two spare units to improve the reliability of supplying firewater to the distribution system. This arrangement is demanded because of the possibility that the main pumping system is removed or placed out of service for maintenance, one pumping system fails on demand or is rendered incapacitated by the fire incident and the final pumping system is then available for service.

NFPA 20 only requires a backup firewater pump where the installed pump is electrically driven and the primary power to it is considered unreliable and secondary power supplies are unavailable.

6.6. BOOSTER FIREWATER PUMPS

Booster firewater pumps are provided where additional pressure is required in the system. They are usually provided where water supply pressures are less than required (although the quantity may be adequate) or where there are extreme elevations in the facility being protected beyond the capabilities of the regular firewater pumps (e.g. at high rise buildings). The booster pump(s) supplies the additional head to provide the water pressure required at the location. In fact, all firewater pumps taking suction from public water supplies or aboveground storage tanks may be considered as booster pumps. Booster firewater pumps should meet the same design and installation requirements as required for regular firewater pumps. Mobile firewater trucks or even fireboats, where suitable arrangements have been provided, may also serve to boost the pressure of the firewater main in an emergency.

Mobile offshore installations (i.e. semi-submersibles) normally have main firewater pumps in their submerged pontoons with booster pumps "topside" due to the elevations involved and demand required. Some offshore installations have also provided booster pumps on derricks or extendible structures from the facility to aid in firefighting capabilities. A booster pump may be required at a shipping dock to ensure the firewater supplied to the vessel though an international shore connection is adequate to meet the requirements of the ship.

Firewater systems are normally designed to operate at about 1,050 kPa (150 psi). If booster firewater pumps are provided that significantly raise the system firewater pressure, special heavy duty or class piping is needed to withstand these higher pressures. Pressure reducing orifices may also be needed at takeoff points for hand held hoses or similar equipment to allow these devices to be used safely. The pressure produced from firewater pumps should never exceed the maximum allowable pressure specified for a firewater system.

6.7. WATER MIST FIREWATER PUMPS

Water mist systems are now commonly used in the design of firewater suppression systems for unique hazards. These systems use very high pressure with small orifice nozzles to produce a highly atomized water mist for the protection of areas that were previously provided with Halon or other fire suppression systems that could affect the environment. Water pumps may be used in these systems to produce low, intermediate or high pressures to assist in the atomizing of the water droplets. These pumps act similarly to a water pressure booster pump.

6.8. JOCKEY PUMPS

Small capacity pumps, commonly referred to as "jockey" pumps, are provided on a firewater system to maintain a constant set pressure on the system, to compensate for small leakages and incidental first aid water stage without the main firewater pump(s) startup or continuously cycling on and off (see Figure 6-5). They are commonly of the centrifugal type, although vane and positive displacement types may be encountered.

Figure 6-5 Fire Pump and Jockey Pump Installation

Most jockey pumps are usually set to start after some flow of water has dropped the normal pressure of fire main to a preset pressure level. Typically, this set point is approximately 35 kPa to 105 kPa (5 to 15 psi) above the startup pressure of the main firewater pump(s), with seven kPa (10 psi) generally selected. Jockey pumps are commonly provided with an electric drive motor and have a relatively small l/min (gpm) capacity compared with the main firewater pumps. Sizing can be accomplished by examining the allowable leakages in NFPA 24, *Installation of Private Fire Service Mains and Their Appurtenances*, for fire mains and ensuring the jockey pump itself will not be continually cycling on and off. Systems without jockey pumps should have a method of protection against the surge of the main firewater pump startup.

Jockey pumps do not require the same standard of testing, integrity or reliability as that required for the main firewater pump(s). For this reason, they should not be credited for firewater supply when calculating available firewater pumping capacity for an installation.

In some cases, a crossover from the utility water system can be used in place of a jockey pump. However, a listed backflow preventer, i.e. check valve is installed to prevent drain down of the firewater system by the utility water system when it is not in service or has bled down. In this case, a designated utility water pump acts as the jockey pump.

The source and capacity of jockey pump water supply should be adequate to meet its demands. It is customary for it to be connected to the same source of supply as the main and backup firewater pumps. Alternatively, it may come from a utility water source where the drawdown of the firewater supply may not be desired.

API Standard 610, *Centrifugal Pumps for Refinery Service* or the Hydraulics Institute, *Standards for Centrifugal, Rotary and Reciprocating Pumps*, may be referenced for the design, construction, installation and performance of jockey pumps supporting firewater systems.

6.9. FIREWATER CIRCULATION PUMPS

In some locations, there may be a need to continually circulate firewater through the system, generally as an additional precautionary measure against water freezing in the distribution network. If the quantity needed to circulate is generally small, the jockey pump may be selected to constantly provide firewater to the system and automatic bleed points can be provided (i.e. relief valves can be provided on the end points of the system which are set to operate at a lower pressure than the shutoff pressure of the jockey pump). Alternatively, separate firewater circulatory pumps can be provided to constantly circulate or recycle water within the system. These pumps are generally common industrial water pumps sized to meet the demand needed for water circulation. Instrumentation is provided on the system to alarm when there is a failure of the water circulatory system. These pumps are not normally required to be listed or approved. In general, this method of protection against water freezing is not preferred, and passive rather than active methods are recommended (i.e. thermal insulation, burial, etc.).

6.10. FOAM PUMPS

Pumps are used as one of the standard methods to provide foam concentrate to foam proportioning devices for fire-fighting purposes (the other method is by a "Bladder" tank), see Figures 6-6 and 6-7. Foam pumps are generally of the positive displacement type because of the high viscosities of foam concentrations. Rotary gear pumps are normally used, which are driven by an electric motor, diesel engine or even a water turbine motor. API Standard 676, *Rotary Pumps for Refinery Service* and the Hydraulics Institute, *Standards for Centrifugal, Rotary and Reciprocating Pumps* lists requirements for the satisfactory design, construction and performance of these types of pumps. Foam pumps should be installed in accordance with the local or national requirements and typically NFPA 11, *Low Expansion Foam and Combined Agent Systems*, is referenced.

Positive displacement (i.e. rotary gear) pumps are generally not made in capacities above 1,900 l/min (500 gpm). In circumstances where foam flow rates are above this level and are required by a single pump, a centrifugal pump is used. Centrifugal pumps, which pump foam concentrates, may experience "slippage" because of the high viscosity of the foam concentrate.

Foam pumps used for firefighting purposes should be considered to be just as critical as firewater pumps. Where a pump is used to supply foam for firefighting, a backup foam pump is provided. Jockey foam pumps are also usually provided if a large foam distribution network is in use. If the support systems to the foam pumps are vulnerable to interruption in an incident or failure (e.g. power to electric motors), other self-contained means with high integrity should be considered.

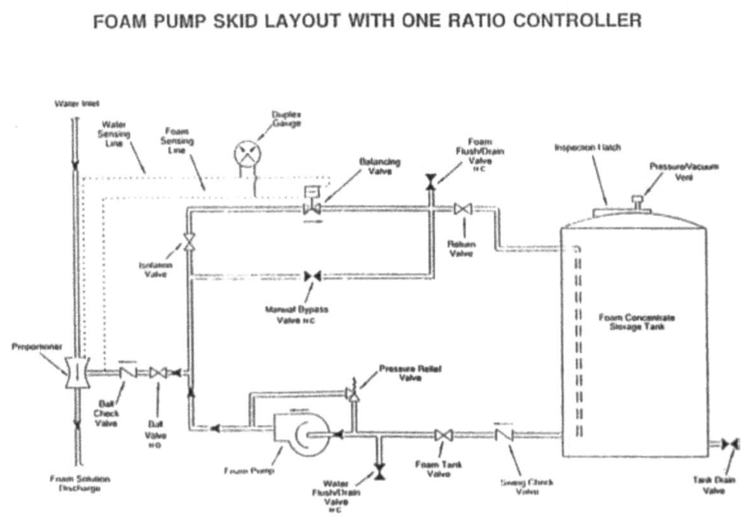

Figure 6-6 Foam Pump Installation Schematic

Source: Courtesy of Chemguard, Inc. Reprinted with permission.

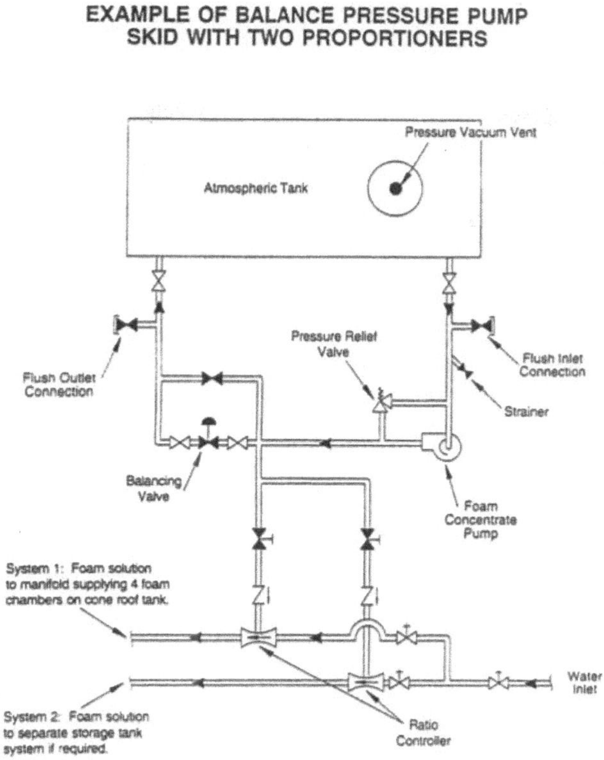

Figure 6-7 Foam Pump Installation Schematic
Source: Courtesy of Chemguard, Inc. Reprinted with permission.

A foam pumping system is considered less reliable than a foam Bladder Tank installation, because of the numerous components associated with a pumping system, so reliance on a Bladder Tank foam system is preferred over a fixed foam pumping system. Foam pumping systems are generally provided where a high pressure is required at the injection of the foam concentrate, which cannot be met by other methods, i.e. a Bladder Tank system.

6.11. PACKAGED AND SKID UNITS

Most firewater pumps are typically provided on a prefabricated skid or as an integral installation package. This is primarily done in order to decrease manufacturing, engineering and the onsite man-hour labor installation costs and time required to install firewater pumps. All the major components are mounted on a structural steel, framed platform that can be shipped in and lifted easily into place onsite. The pump, driver, controller and some supplemental accessories are usually pre-assembled and mounted on the skid (see Figure 6-8).

Figure 6-8 Fire Pump Skid Unit
Source: Photo Courtesy of Patterson Pump Ireland Limited.

The skid installation also helps avoid major alignment problems between the driver and the pump or right angle gear drive, although all pumps need alignment confirmation after initial installation, and the alignment needs to be rechecked periodically. Pump alignments should be in accordance with the Hydraulics Institute, *Standard for Centrifugal, Rotary and Reciprocating Pumps.*

Where an enclosure for the firewater pumping system is required, a modular system can be provided with its own totally enclosed housing unit. Because all the components are provided by one supplier, the responsibility for correct installation and operation is reduced to one single point of contact.

Standardized skid arrangements can be utilized or custom design skids are constructed as specified by the purchaser during the initial design specification. Whether standard or customized skids are used, size, orientation and interfaces to the onsite facilities must be confirmed. These include pump suction and discharge directions, control and instrument cable routings, controller visibility and access, fuel or power connections or refill arrangements, allowable skid size, skid tie-down points, drain ports and sewer connections, and testing or maintenance accessibility concerns. Where the skid is to be lifted into place with slings, the sling lifting points should be specifically identified and potential interference during the lifting operation with equipment on the skid should be considered. If the unit is also supplied with its own enclosure, this should meet the fire and explosion impacts that the firewater pump site may be exposed to. Retrofit provisions may require the skid to actually be supplied in sections in order for it to be fitted into the space allocated for the new pump installation.

6.12. RETROFIT IMPROVEMENTS TO EXISTING FIREWATER PUMPS

Some occasions may occur that require an increase of the amount of firewater provided to a facility from an existing firewater pumping system. The most common practice is to provide additional firewater pumps to the facility to achieve this requirement. A commonly employed innovative and economical approach is to replace or re-machine only the pump impeller or provide an improved hydraulic end (in a lineshaft pump) to a higher capacity rating to achieve the required firewater flow. In fact, API Standard 610, *Centrifugal Pumps for General Refinery Service*, states pumps should be capable of a five percent head increase at rated conditions by installation of a new impeller or impellers. This approach is helpful if the available space to install a new pump is limited.

This approach is only possible if the pump driver and drive transmission system is capable of the increased demands of the higher water flow and can withstand the resulting increase in stress in the system. The complete existing system must be studied and examined by experienced pump suppliers in order to determine the validity of such a proposal (e.g. Torsional Vibration Analysis (TVA) analysis). Where this is feasible, improvements in the order of 30 percent of performance may be realized.

6.13. FUTURE EXPANSION

Provision and arrangement of firewater pumps should consider future expansion of modification of the facility. When such expansion may be feasible, available space, tie-in connections, line sizing and hydraulics of the system should be evaluated.

6.14. RELIANCE ON MOBILE FIREWATER PUMPING APPARATUS

In some cases, the proximity of a local fire station or provision of a dedicated fire station within a large industrial complex can be relied upon to provide backup firewater

Figure 6-9 Mobile Fire Apparatus

pumping capability to the system. In fact, historical evidence indicates that when the fixed fire pumps have been impacted by a major fire or explosion incident, mobile fire apparatus has to be heavily relied upon as a backup mechanism. Previous coordination with the fire station as to their capabilities, mobile apparatus accessibility, connection points, drafting sites, emergency admittance and manpower should be evaluated and incorporated into emergency pre-fire plans for the facility (see Figure 6-9).

Firewater pumper stub-up connections can be strategically positioned on the firewater distribution network to supplement the fixed supply sources or boost the firewater main pressure.

6.15. PORTABLE PUMPS

Small skid mounted firewater pumps are available, which are designed to be easily transportable and operated rapidly for emergency conditions. Some of these pumps are designed to be placed in a water supply and float on its surface, even in conditions when there is only 15.3 cm (6 inches) of water depth.

6.16. NFPA 20 VERSUS API 610 AND OTHER PUMP TYPES

Often, a pump constructed to API Standard 610, *Centrifugal Pumps for General Refinery Service*, or other similar pump specifications may be suggested in place of a NFPA 20 or other listed pumps due to economical concerns or availability of company internal spares. The main difference between these two pumps is the NPFA pumps have been rigorously tested to an independent test laboratory specification, the specific pump curve required by NFPA or other regulatory codes and some minor pump construction details. Additionally, NFPA requires specifically approved accessories to be provided and prescribes certain pump and piping installation details. Without these provisions, the availability and reliability of the firewater pump to deliver water on demand is decreased. Due to the importance of firewater pumps in the protection of multi-million or even billion dollar facilities, the tradeoff for pump reliability for non-standard firewater pumps is not justified.

Facilities that are less critical, unmanned or have low economical value may consider the use of firewater pumps that do not meet NFPA requirements, as their loss may be of little consequence. In addition, the protection which the non-standard pump installation offers is an improvement over the previously unavailable firewater supplies, since if the non-standard pump is not provided no firewater will be available to the installation in these cases.

7 PUMP INSTALLATION, PIPING ARRANGEMENTS AND ACCESSORIES

7.1. INTRODUCTION

In general because of their critical support in life safety and property protection, firewater pumping system components are normally required to be listed, approved or certified by the applicable fire code for the facility and installed as per a recognized fire code. Code requirements are obtained during the design and installation of the system. It is normally felt that if the pump installation is built to "code", it is acceptable and fully reliable. If such were the case, failure of a firewater pump due to design or installation deficiencies should never occur. The manufacturer's listing or approval is obtained from a recognized testing organization concerned with their design and operation to a given set of parameters. This approval is generally specified for the pump, driver, control system and some of the associated water handling equipment.

7.2. CODE REQUIREMENTS

Worldwide, in countries with mostly western industrial influence, two major sets of fire codes (or local codes based on these) can be generally encountered—American or European. The American code is defined by the National Fire Protection Association (NFPA) Fire Codes and the European by the Comite European des Assurances (CEA) Rules. NFPA is used in North America and those countries with an historical connection with the United States (e.g. The Philippines) and at worldwide oil or chemical plants where American design standards have dominated the design of the facility due to the multi-national company involvement with the installation and its standards (e.g. Saudi Arabia). CEA rules are used in Europe and in countries with a historical European connection (e.g. Australia). Currently, most countries in Europe also have their own rules with each specifying slightly different requirements.

Conformance to NFPA or CEA requirements is almost universally accomplished by obtaining an independent evaluation of the equipment manufacture and performance by a recognized testing agency. Underwriters' Laboratories, Inc. (UL) or Factory Mutual (FM) listing or approval is obtained for American based systems. Similar European agencies are used for CEA requirements (e.g. Loss Prevention Certification Board (LPCB) is used in the United Kingdom). Approval agencies perform various inspections and tests of the equipment according to their own approval requirements. These tests generally concern bolt and shaft stresses, bearing life, materials of construction, parts lists and hydrostatic and performance tests. Successful evaluation by the testing organization designates it to be listed or approved by its organization. These

Fire Fighting Pumping Systems at Industrial Facilities. DOI: 10.1016/B978-1-4377-4471-2.00007-4

organizations publish periodic directories of products that have been listed or approved.

Today, most codes are prescriptive in nature, i.e. they set specific detailed equipment and installation requirements, rather than meet design performance and reliability. There is a trend for fire codes to become more performance-based than prescriptive. No doubt this trend will affect the requirements for firewater pumping systems in the future. There are also installations where unique requirements have allowed the installation of a firewater pumping system that did not meet the detailed prescription requirements of the fire codes, but instead its objective requirements have been met, thereby allowing it to be acceptable to the authority having jurisdiction (AHJ).

NFPA 20, *Standard for the Installation of Centrifugal Firewater Pumps*, is recognized worldwide for the provision of firewater pumps. NFPA is a non-profit, technical and educational association without any governmental affiliation. It has no enforcement authority and only when NFPA codes are "adopted" by an organization that has jurisdiction over the facility are its requirements applicable by law. All NFPA Fire Codes are arrived at by a "consensus". Their contents have been arrived at through committee action and general agreement from a wide range of individuals from various industries, manufacturers and governmental bodies. In some fashion, they may be thought of as arbitrary as they may be based on a few selected incidents without a full risk evaluation of the basis of the requirement. They are generally thought to represent the minimum acceptable standards for fire protection requirements.

Individual company requirements may be below or above NFPA codes in certain instances, because some portion of the fire code cannot be applied practically in its own environment. The AHJ should realize and accept the limitations of universal application of all NFPA codes. Additionally, insurance guidelines are generally more protective than NFPA fire codes due to the greater "private" concern (i.e. by the insurer) for economic gain and prevention of losses.

It should also be realized that just because a fire pump installation meets all the fire code requirements, it does not guarantee its ability to supply water in an emergency. Some aspects of how the pump is designed and installed may still cause it to be unreliable or subject to detrimental exposures. The code often does not specify the hazard(s) they are intended to protect against. In some cases the fire code is very generic in its approach to requirements for the installation of a fire pump, e.g. protection to be provided from plant fire and explosion hazards, and these are left entirely up to the facility owner to determine what the adequate requirements are.

Local building codes also have to be reviewed to ascertain the safety features required. Some localities prohibit the direct connection of private firewater pumping systems to the public water mains because of health or hydraulic performance concerns regarding their systems. The placement of diesel engines as pump drivers, in buildings also has to meet the building code requirements for these installations. It is therefore highly incumbent upon the individual installing a firewater pump to ensure all aspects are evaluated, not just those listed in one standard fire code (i.e. NFPA 20).

7.3. LISTING REQUIREMENTS

Most fire codes have a listing or approval requirement for the provision of the firewater pumping system components. Listing or approval implies that independent testing or verification of the performance of a device to a recognized standard has occurred. The main purpose of listing or approvals is to provide some assurance that the installed system will meet performance and reliability requirements. Components that are installed which are required to be listed may not have the same operational reliability factor that a listed or approved component may have.

Interestingly, not all equipment used for firewater pumping systems is available or manufactured with a listing or approval. However, these unlisted or unapproved devices may still have a very high reliability factor. For example, right angle gear drives are not required to be approved as per the NFPA code, nor are they specifically required to meet any other standard for firewater service (although some manufacturers are now providing the feature through one of the approval agencies which offers a listing for it, i.e. UL; FM). Yet, they are routinely used as power transmission components for vertical lineshaft firewater pumps and are fully recognized by fire codes for this purpose. Table 7-1 provides a listing of items requiring approval or listing required by NFPA 20 (2010).

Table 7-1 Fire Pump Components to be Listed per NFPA

Item	NFPA 20 (2010) Paragraph
Centrifugal Fire Pump	4.7.1
Positive Displacement Pump	8.1.2.1
Automatic Relief Valve	4.11.1.1
Suction OS & Y Valve	4.14.5.1
Discharge Check Valve	4.15.6
Discharge Indicating Gate Valve	4.15.7
Discharge Indicating Butterfly Valve	4.15.7
Metering Device or Fixed Testing Nozzle	4.20.2.1.1
Hose Valve	4.20.3.1.1
Backflow Prevention Device	4.27.1
Automatic Air Release	6.3.3.1
Flexible Coupling or Connecting Shaft	6.5.1.2 and 7.5.1.8.1
Electric Motor Driver	9.5.1.1
Electric Motor Controller	10.1.2.1
Power Transfer Switch	10.1.2.1
Diesel Engine Driver	11.2.1
Battery Contactors	11.2.7.2.2.1 (a)
Battery Charger	12.6 (1)
Cooling water automatic valve (for diesel engine service)	11.2.8.5.3.5
Flexible Fuel Hose	11.4.4.1
Diesel Engine Controller	12.1.3.1

Table 7-2 UL Test Standards for Listed Devices

Item	UL Standard
Centrifugal Fire Pump	UL 448
Positive Displacement Pumps	UL 448C
Fire Pump Relief Valves	UL 1478
Suction OS & Y Valve	UL 262
Discharge Check Valve	UL 312
Discharge Indicating Gate Valve	UL 262
Discharge Indicating Butterfly Valve	UL 1091
Metering Device or Fixed Testing Nozzle	UL 385
Hose Valve	UL 668
Backflow Prevention Device	UL 312
Automatic Air Release	n/a*
Flexible Coupling or Connecting Shaft	UL 448A
Fire Pump Motors	UL 1004-5
Fire Pump Controller	UL 218
Power Transfer Switch	UL 1008
Diesel Engine Driver	UL 1247
Battery Charger	UL 1236
Flexible Fuel Hose	UL 536
Diesel Engine Controller	UL 218

*n/a—not available.

In many cases, for industrial usage the use of non-listed or approved equipment is allowed. Some reasons for this are as follows:

- Common use within a particular class of industry for the device in firewater service.
- The manufacturer states it is of equivalent construction, reliability and designed for firewater use according to recognized codes.
- The required equipment is not manufactured or easily procured that is to be listed.
- Other listed alternatives are not practical for the installation.
- A particular non-listed component may be a more practical application than the listed component.

Many large industrial risks require, and use, fire pumps with capacities that are far above those tested by recognized laboratories but otherwise would meet the listing requirements. The final decision for acceptability rests with the AHJ (see Table 7-2).

7.4. TYPICAL INSTALLATION

As a minimum, firewater pumps and their installations should conform to local or National Authority regulatory requirements (e.g. NFPA 20, Loss Prevention Council (UK) (LPC), etc.) which are necessary for legal compliance. In addition, some insurance underwriters require fire pump installations to be approved by their offices. A company may also have its own policies, standards or specifications for the installation of firewater pumps that will be audited against during internal inspection.

Typically, firewater pumps are rated in standard capacities of 1,892, 2,839, 3,785, 5,677, 7,570, 9,462 l/min (500, 750, 1,000, 1,500, 2,000 and 2,500 gpm). Pump discharge outlets are sized according to the listed rating of the pump. Larger capacities and special arrangements can be used in specially engineered systems subject to approval from the local authority.

Pump rated output pressures can also vary. Most sprinkler or water spray systems require between 350 to 700 kPa (50 to 100 psi). Firewater monitors and foam systems generally require higher pressures in the range of 700 to 1,050 kPa (100 to 150 psi) to be highly effective. Mobile fire pumping apparatus can be supplied from a water distribution network with pressures of 140 to 350 kPa (20 to 50 psi). In general, fixed fire pumps are specified with net discharge pressures in the range of 875 to 1,050 kPa (125 to 150 psi). Firewater pumps with discharge pressures greater than this would generally produce shutoff pressures exceeding the working pressures of most of the fire protection equipment, generally 1,225 kPa (175 psi), and would require the addition of pressure regulation valves.

Figure 7-1 depicts common firewater pump and piping arrangements for horizontally driven firewater pumps taking suction from a firewater storage tank. Depending on whether a horizontal or vertical shaft driver is chosen, certain advantages and disadvantage may be incurred, see Table 7-3.

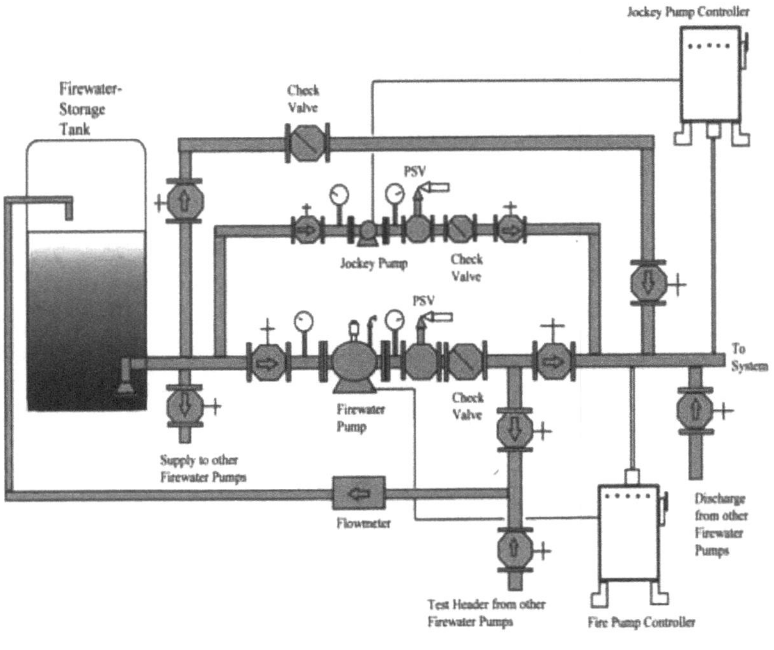

Figure 7-1 Typical Pump Piping Arrangements

Table 7-3 Comparison of Horizontal Split Case to Vertical Pumps

Feature	Horizontal Split Case	Vertical Split Case
Floor space	More than Vertical	Less than Horizontal
Exposure to flooding conditions in pump area	More than Vertical	Less than Horizontal
Inline piping arrangement	Has to be specified	Allows piping in any arrangement

7.5. LOCATION AND SEPARATION FROM PROCESS AREAS

Analysis of fire losses for chemical plants, refineries and other similar petroleum installations indicates that about 10 percent of the incidents involved impairment of the firewater pumping systems at the time of the incident (see Appendix A.1). The losses that resulted because of firewater pump failure suggest the need to carefully examine the location chosen for the pump installation in high hazard industrial process areas.

The primary concern in the reliability of a firewater system during an incident is that the pump and its supply system are not to be affected by the incident. Explosions and high hazard hydrocarbon and chemical fires are the primary incidents that mainly affect firewater pump systems. Explosions in particular have been shown to be particularly devastating to the firewater supply system where their arrangement, pipework or location has been vulnerable to an explosion event or its effects. Realizing there may be different categories of explosions, the possibility of an Unconfined Vapor Cloud Explosion (UVCE) is of most concern.

UVCEs primarily occur where large volumes of flammable or combustible gases may be released primarily in areas with some congestion. These installations tend to be gas and chemical plants, refineries, gas storage, pipeline compressor stations and offshore installations that handle hydrocarbons. Firewater pumping systems fail most often when a vapor cloud explosion occurs. In a review of the 100 largest losses in oil and chemical plants, it was discovered that steam and electrical utilities are particularly susceptible to damage from explosions. It has been found that in 92 percent of the cases where firewater pumping systems have failed, they were driven by electric motors or steam turbines. Because of the high levels of these power system impairments at industrial facilities, one insurance underwriter recommends that at least 50 percent of the drivers for firewater pumps be provided as diesel engines.

Typical spacing distances recommended by insurance underwriters for fire pumps from hydrocarbon or chemical process or storage areas is 107 to 61 meters (350 to 200 ft.) (see Table 7-4). Major integrated petroleum companies on average recommend approximately 61 to 51 meters (200 to 167 ft.) for firewater pumps from process areas or storage areas. As can easily be seen, a major disparity exists between the

Table 7-4 Common Spacing Distances for Firewater Pumps

Hazard	Spacing Distance (Insurance Industry)	Spacing Distance (Oil Industry*)
Process Area (high hazard)	91 m (300 ft.)	61 m (200 ft.)
Process Area (medium hazard)	91 m (300 ft.)	61 m (200 ft.)
Process Area (low hazard)	61 m (200 ft.)	51 m (167 ft.)
Storage Tanks (hydrocarbons)	107 m (350 ft.)	61 m (200 ft.)
Office and Utility Buildings	15 m (50 ft.)	7.5 m (25 ft.)

*Average from spacing guides for six major integrated oil companies, circa 1990s.

insurance industry and the operators of the properties they insure for the amount of spacing to be provided to firewater pumps. This has probably led to some instances where the firewater pumps have been impacted by the incident they were intended to protect against. Each side (insurance industry versus facility owner) has a vested interest in minimizing economic impacts, however, the "operators" have apparently viewed the larger facilities needed for increased areas for spacing as an uneconomical need during initial design rather than a long term benefit against possible process incidents.

The location chosen for the placement of a firewater pump is usually dependent on the source of supply. It is highly advantageous, especially in facilities where there is the possibility of catastrophic destruction, to have pumps situated in different parts of the facility. This will prevent the exposure of all pumping capability to a simultaneous loss. It will also have improved hydraulic effects for the system. Even when spacing distances are met, masonry construction should be provided for firewater pump houses and open sided shelters are only recommended for remote locations. They should be spaced away from other buildings or structures to prevent impact from their collapse or fire exposure.

Firewater pumps should be located at a higher elevation of the facility, and upwind from the process areas where practical. Low areas and downwind locations may be areas where combustible gases can accumulate and impact on the firewater pump operation. Low areas may also be susceptible to spill collection or flooding either from process materials or rainwater. If an upwind location is not available, a crosswind site should be chosen as the second alternative. The predominant wind direction for a location should be obtained from a tabulated wind rose for the site compiled from historical data.

For offshore installations, meeting some of these spacing distances is obviously wholly uneconomical. The best location for a firewater pump offshore is on a separate utility or accommodation platform when these are provided. Otherwise, they should be in a utility or non-process module that is not located adjacent to a hydrocarbon handling area. All firewater pumps located on hydrocarbon processing platforms should be examined to determine their vulnerability to explosion incidents. Additionally, offshore petroleum installation regulations in the North Sea require that firewater pumps must be provided separately from each other (i.e., the pumps or pump rooms should not be next to each other). They should never be located adjacent to an area that handles combustible or flammable materials. During the Piper Alpha offshore fire incident in 1988, it was speculated that one of the main initial explosions impacted the two facility firewater pumps because the utility module in which they were located was next to the gas compressor module. The wall separating the compression module from the utility module, although it was a fire-rated wall, had little to no blast resistance. Additionally these two firewater pumps were located relatively close to each other, separated only by a firewall and adjacent to a common utility area. Firewater pump rooms or buildings should be separated from other non-process areas by a firewall where they share a common wall.

7.6. PUMP SEPARATION

To avoid common failure incidents, main, backup and other supporting firewater pumps, such as booster pumps should not be located immediately next to each other and ideally should be housed in separate locations at the facility.

They should feed into the firewater distribution system at points that are as remote from each other as practical. In practical applications, except in offshore installations, most small and medium facilities have a single firewater storage tank requiring the

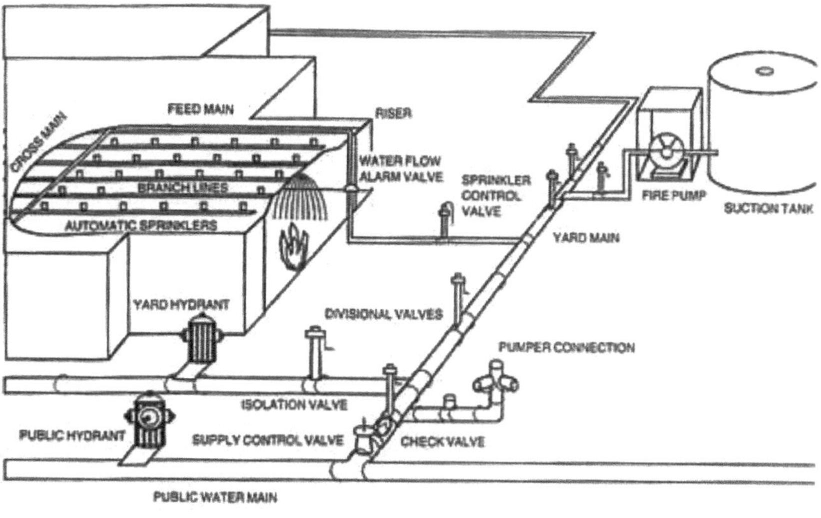

Figure 7-2 Typical Arrangement of Several Firewater Supplies at an Industrial Facility
Source: Copyright Factory Mutual Research Corporation, reprinted with permission.

siting of all firewater pumps close to it. Even in this situation it may be wise to provide separate tie-in points from the main and backup pumps to the firewater distribution system. Again, this mostly depends on the hazard the facility represents and the distance the firewater pumps are located from the process or storage areas. A further advantage of locating firewater pumps at separate and remote tie-in points is the enhanced hydraulic pressures available to the firewater distribution system. Water supplies from other alternative sources, such as public water mains, mobile firewater pumping apparatus, should again be located separately from each so that an impact on one source will not affect another (see Figure 7-2).

In some instances, a firewater pump may be required to be placed in an electrically classified location. Although, it is best to avoid this, the constraints of space, weight distribution, and separation requirements for offshore facilities may preclude any other option. Requirements for the use of a firewater pump in electrically classified locations are discussed in Chapter 12.

7.7. PUMP ROOM OR BUILDING CONSTRUCTION

When fire pumps are provided in buildings, they should be of masonry construction with non-combustible roofs. Open-sided shelters should be avoided except in remote locations. Pump rooms that share a common wall with other occupancies should evaluate the level of fire resistance and even blast resistance the wall may need to provide. Adequate air supplies should be provided. These enclosures should be provided with access doors to allow the complete removal of all major assemblies of the pumping system (driver, gearbox, pump, etc.). Where overhead hoist beams have not been provided for the handling of heavy assemblies, the access of portable lifting equipment into the pump area should be incorporated.

7.8. SPECIAL LOCATIONS

7.8.1. OFFSHORE FACILITIES

Most firewater pumps for offshore structures are lineshaft vertical turbine or electro-submersible centrifugal types, which direct water from sea level to the topsides via a column pipe. The design installation of these pumps is in a case almost by themselves as compared to other firewater pumps due to their complexity, support arrangements, vulnerability to incidents and economics of the installation. In fact, most international or national fire codes do not specifically address such installations and unfortunately little public literature is available on their installation requirements. The obvious concern regarding the pump location relative to hydrocarbon explosions and fires is of utmost importance. Therefore, the provided firewater pumps should be located as far as is practical from the potential source of these incidents (see Figure 7-3). Additionally, the space requirement and weight of offshore pumps is of concern because it directly affects the size and load-bearing capacity of the platform it rests on. Because the weight of all equipment in offshore facilities must be supported on a

Figure 7-3 Norwegian North Sea Oil Platform

dedicated platform, the cost of the platform increases when additional space and weight is required. Therefore, considerable economical pressures exist to limit space and weight requirements during the design of these facilities.

The offshore pump lift column should be protected against wave action and mechanical damage. Workboats routinely visit these structures and have occasionally been known to bump into the structure because of unanticipated wave or current action. Additionally, when lineshaft pumps are used, the column has to be positioned absolutely vertical to avoid alignment and stress failures in the pump assembly. Common practice is to place the pump column in a caisson and locate it where it will be protected by the structural steel framing of the platform. If a concrete support structure is available it is placed inside this. In both instances it minimizes the potential damage from a collision with the platform from a marine vessel.

A location to remove or lift a firewater pump or its driver by the platform crane or other lifting mechanism should be incorporated. Headroom must also be provided at the location of the pump column in order for the pump column pipes to be physically capable of removal during pump maintenance or replacement.

There may also be instances where two separate offshore installations are required to function as a single facility. Many existing fixed offshore oil and gas production platforms periodically require the services of a mobile drilling rig to drill new wells or provide workover services to existing wells. In these cases, the mobile rig is positioned over and on top of the fixed platform to form an integrated unit. There may be a need to interconnect the separate facility firewater systems to improve the reliability and supplies of both during the drilling or workover operations. Hose connections between the two systems can be provided with breakaway connections. This allows the provision of one system to supplement the capability of the other system and improve the overall protection of the integrated unit. The provision of breakaway couplings allows the rig to quickly move off during an emergency without impact to its or the platform's firewater system.

7.8.2. ARCTIC LOCATIONS

Low temperature or freezing conditions are the highest concern where firewater pumps are provided in arctic conditions. Cooling water and lubrication oil immersion heaters are mandatory in these conditions for both diesel engines and gear drives. Protection of the waterlines from freezing by depth of burial, insulation and circulation means is commonly employed. Water storage tanks are commonly heated and insulated and firewater pumps are located in a heated pump house. Disposal of firewater during a fire incident in these locations may also be of a concern if left in the open, causing major ice build-up.

7.8.3. ARID LOCATIONS

In arid locations, extreme heat, constant direct sunlight and sandstorms may be prevalent. Precautions must be taken against these conditions to prevent damage or early failure of the pumping system. Buildings or sun shades are provided over equipment. Critical examinations are made to systems or components that can fail due to contamination with dusts or sand, overheating or rapid deterioration of "rubber" or "plastic" components because of prolonged exposure to elevated temperatures or sunlight radiation (i.e. seals, drive belts, etc.) causing them to lose their elasticity. Air intakes for internal combustible engines are orientated to face downwind to lessen the probability of sand and dust ingestion. Suitable heavy-duty air filters or sand baffles are commonly provided.

7.8.4. TROPICAL LOCATIONS

Heavy rains (monsoons, hurricanes, typhoons, etc.), animal or insect infestations and direct sunlight exposures are the most common concerns for firewater pumps located in tropical locations. Heavy rains can produce flooding conditions that may envelop the pump location, especially if it is taking suction from a river source, without flood control measures. Elevated locations should be considered in these cases. Heavy rains are usually accompanied by high winds that can carry objects which can damage a pump installation.

Insect or rodent infestations can cause blockages in vents or contamination of fuel systems and also deterioration of soft materials. Frequent inspections and suitable screens should be considered. Direct sunlight can damage rubber components and reduce their elasticity. Pumps houses are provided to protect against the rain, winds, sunlight and animal disturbances.

7.8.5. EARTHQUAKE ZONES

Firewater pumps located in areas susceptible to earthquakes should be provided with suitable restraints. The extent of these restraints are normally dictated by local ordinances and primarily concern the bracing of pump pipework and adequate securing of pump base plates and controller panels for earthquake forces. Pumphouses should be adequately constructed and braced so they will not collapse onto the firewater pump or distribution piping.

7.9. MULTIPLE PUMP INSTALLATIONS

Multiple pumps in a single location should have a symmetrical piping layout to achieve similar flow conditions to and from all pumps. A common panel should be available to determine the status of all pumps at a single glance. Individual pumps should be able to be isolated or tested without affecting the use of other pumps. Use of common headers and support systems should be provided to improve the economics and hydraulic flow of the entire system.

Where multiple pumps are provided, efforts should be made to purchase identical units to simplify familiarity, spare parts, maintenance and design information. Often where the multiple purchases of identical units are involved, a manufacturer can offer a discount in costs due to overall lower manufacturing costs per unit for an order of multiple identical units.

7.10. PUMP ROTATION

The rotation of the fire pump (clockwise or counter-clockwise) is important because it defines the location of the suction and discharge of a horizontally arranged pumping system. Normal design is to have pump rotation as clockwise (CW). Counter-clockwise (CCW) rotational drivers are available on special order at additional cost and possibly with an increase in the manufacturer's fabrication schedule.

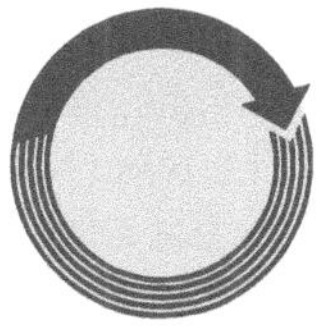

7.11. RELIEF VALVES

A pressure *safety* relief valve (PSV or RV) is provided to protect the pump and its associated pump from the unexpected overpressure of the system from the pumping unit (see Figures 7-4 and 7-5). Overpressures may result due to high speed or over-revving of the pump driver or backpressure onto the system. A relief valve is required whenever the pump shutoff pressure plus the suction pressure exceeds the

Figure 7-4 Relief Valve Fitted to Pump Discharge
Source: Photo Courtesy Patterson Pump.

pressure rating of the system components. Relief valves should not be provided to correct a situation where constant speed motor driven pumps have been incorrectly provided and their discharge pressure exceeds the pressure rating of the system piping.

The relief valve should be installed in the piping connected to the discharge of the pump prior to the discharge check valve. The discharge of the relief valve outlet should not impinge the fire pump system, on other equipment or locations where personnel could be present. Normally it should discharge into the open so that correct or incorrect operation can be verified. The discharge is generally piped to an open drain funnel or cone to the facility gravity sewer. Whenever the relief valve does not discharge to the atmosphere, a sight flow indicator (e.g. sight glass) should be provided to determine if the valve is passing or has operated. The relief valve should not be piped back to the suction side of the pump. NFPA 20 does not allow an isolation valve to be provided in the suction or discharge of the relief valve piping due to the possibility that the valves may be accidentally closed and the pumping system left without overpressure protection. Some locations have chosen to provide isolation valves that are locked in the open position, and therefore when change-out or repairs are needed of the installed valve it can be readily replaced, in a similar manner that isolation valves are provided to protect process vessels.

The relief valve should be set according to the company's standards for the protection of equipment for safety relief valves. For firewater pumps it is normally set slightly above the pump's rated shutoff (no water flow) pressure to relieve water during overpressure or overspeed conditions.

Relief valves need to be recalibrated periodically. The frequency of recalibration should be determined by the policy adopted by the facility for relief valve

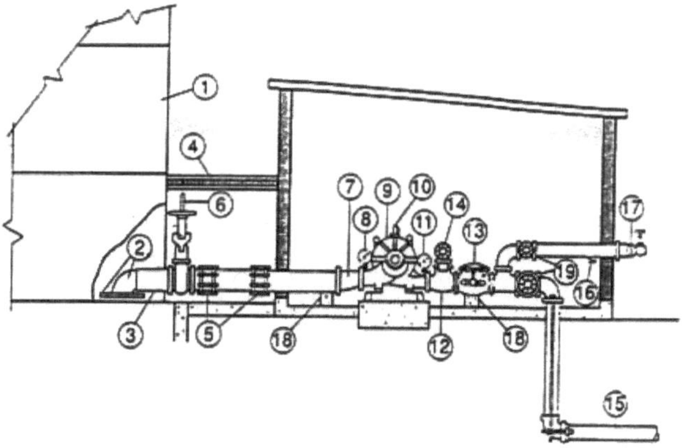

1. Aboveground suction tank
2. Entrance elbow and 4 ft × 4 ft (1.2 m × 1.2 m) square vortex plate. Distance above bottom of tank–one-half diameter of suction pipe with a minimum of 6 in. (152 mm)
3. Suction pipe
4. Frostproof casing
5. Flexible couplings for strain relief
6. O.S. & Y. gate valve (see 2-9.7 and A-2-9.7)
7. Eccentric reducer
8. Suction gauge
9. Horizontal split-case fire pump
10. Automatic air release
11. Discharge gauge
12. Reducing discharge tee
13. Discharge check valve
14. Relief valve (if required)
15. Discharge pipe
16. Drain valve or ball drip
17. Hose valve manifold with hose valves
18. Pipe supports
19. Indicating gate or indicating butterfly valve

Figure 7-5 NFPA Horizontal Pump Installation from Storage Tank
Source: Reprint with permission from NFPA 20, *Installation of Centrifugal Fire Pumps.* Copyright ©1996, National Fire Protection Association, Quincy, MA 02269. This reprinted material is not the complete and official position of the National Fire Protection Association, on the referenced subject which is represented only by the standard in its entirety.

recertification. A calibration tag should be attached to the valve indicating its calibration date and preferably, its set pressure. Isolation valves should be provided to the relief valve for maintenance or removal and should be normally locked or carsealed in the open position.

7.12. CIRCULATION RELIEF VALVES

All firewater pumps are to be provided with automatic circulation relief valves (CRV) to provide water circulation through the pump when flow to the distribution network is not occurring to prevent the pump from overheating. The set point of the CRV should occur just below the shutoff pressure rating of the pump, with minimum suction pressure taken into account. Engine-driven firewater pumps which use a portion of the pump discharge water for cooling purposes need not be provided with a circulation relief valve.

7.13. PRESSURE AND FLOW CONTROL VALVES

Where limitations are placed on the firewater distribution system for reasons of pressure or flow limitations from a firewater pump, a pressure (PCV) or flow control valve (FCV) is provided. The PCV or FCV limits the pressure or flow delivered to the firewater distribution system. Such restrictions are placed on a system where downstream components, i.e. hose reels or other equipment would be damaged or cannot be controlled manually due to high pressures or flows. Firemain pressures are generally not allowed to exceed 1,050 kPa (150 psi) for these reasons, unless higher rated pipe classes are used and pressure restriction orifices are provided at end devices. The PCV or FCV is provided at the connection point of the firewater pump header into the distribution network.

7.14. ISOLATION VALVES

Manual isolation valves are provided at the firewater pump to permit its removal without impacting the entire system. They are also used to direct water for flow performance testing (see Figures 7-8 and 7-9). All isolation valves should be of the indicating type and NFPA 20 recommends these be listed (e.g. FM or UL).

Indicating valves are those that provide an easily distinguishable visual method to determine the open or closed status of the valve. These valves should be of the wheel and gear type as opposed to a quarter turn ball type to avoid rapid pressure changes during closure causing water hammer.

Outside Stem and Yoke (OS&Y) valves are universally used for exposed valves, from which the visibility of the valve stem either "in" or "out" of the valve body will indicate the "closed" or "open" status, respectively, of the valve at a glance. Valves which are provided on the distribution network from the firewater pump may be buried underground for protection and post-indicator-valves (PIVs) are commonly used. These are provided with position indicator windows showing "shut" or "open" status.

7.15. BYPASS CAPABILITY

Where the source of the firewater pump supply can be used to supply the distribution system with some margin, without the use of the firewater pump (i.e. the pump suction pressure is sufficient to be of fire protection use), a bypass is recommended as stipulated by NFPA 20. The purpose of the bypass is to make the supply source available to the distribution system in the case of total loss of the pumping capability at the facility.

7.16. PRESSURE GAGES

Pressure gages should be provided on the suction and discharge of a firewater pump. A suction pressure gage is not required for vertical shaft, turbine type pumps taking suction from a well, open wet pit or offshore facility (as the gage itself would be in the submerged liquid). These gages provide assistance to verify the normal pump function during daily operations and to assist in the conduction of periodic flow performance verification tests. They should be placed or fitted where they are easily readable during operation of the pump by the personnel that attend the operation of the system and give an accurate reading of the performance of the pumping system. Additionally, a pressure transducer can supplement the gages to provide direct readout of pump pressures in the control room or through the distributed control system (DCS) or other utility monitoring system.

The range of the discharge pressure gage should be able to accommodate the full pressure range of the firewater pump output with range of twice the rated pressure of the pump. Suction pressure gages should have a range of twice the maximum rated suction pressure. Normal operating pressures should be indicated between 30 to 70 percent of the scale range. It should use pressure units that are consistent and standardized for use at the facility. Where a suction lift is involved, a compound gage (i.e. with vacuum reading capability) should be fitted to the intake side of the pump. Gage dials should be at least 89 mm (3 $^1/_2$ inch) in diameter and provided with a 6.25 mm ($^1/_4$ inch) gage valve. Pressure gages with a white background, black numerals and a black point are normally installed. The glass used in gage dials should be made of safety glass in compliance with American National Standards Institute (ANSI) Z26.1 or equal. They should conform to ANSI B40.1, *Gauges - Pressure, Indicating Dial Type - Elastic Element*, or ANSI B40.1M, if metric units are in use. Underwriters Laboratories test gages for Fire Protection Service under their standard UL 393. Gages should be marked in the units that are adopted for the facility (i.e. Metric or English).

Bourdon tube pressure gages are commonly selected for pressure measurement. They have an accuracy of approximately +/- 2 percent of the full-scale deflection. They are susceptible to mechanical shock, pressure surges and the Bourdon tube will also tend to fatigue over time and inevitably, some drift in the readings will occur over time. It is also possible that the tube connection to the gage may become plugged with contaminants. Therefore as a rule, all pressure gages should be calibrated before tests of the firewater pumping system are performed. A calibration sticker should be affixed to the gage indicating the date of the last calibration.

The location chosen for the pressure gage installation may cause it to fluctuate somewhat if there is some turbulence in the water flow in the piping. For a relatively steady pressure reading, a straight flow of piping at the pressure gage is necessary to reduce turbulence.

7.17. PRESSURE RECORDERS

FM, NFPA and UL requirements require water supply pressure recorders; however, many industrial sites today continually observe and record the performance of their utility systems through a DCS or other facility monitoring system. This offers the advantage of real-time call-up of the performance of water systems from a manned console.

7.18. FLOW MEASUREMENT CAPABILITY

A method to accurately measure the quantity of flow produced by a firewater pump during periodic flow performance verification should be provided. Numerous methods are available, including multiple 63.5 mm (2 ½ inch) test outlets on a test header, orifice plates, and a variety of state-of-the-art electronic or magnetic flow meters.

A test header with 63.5 mm (2 ½ inch) outlets for manually undertaking pitot tube flow measuring can be used with reference to hydraulic flow tables. Flow is released from the outlets through fire hoses to smooth bore test nozzles. The laminar flow produced by the smooth bore nozzles is measured via a manually inserted pitot tube for water capacity measurement. The values obtained from the pitot tube can be interpreted for capacity from circular outlets, with correction for a coefficient of roughness of the outlet, through any standard hydraulics reference table (see Chapter 13).

Historically, most installations have used an orifice plate flow meter (see Figure 7-6). This application is easily applied as it slips into the flange connections of the discharge piping. The orifice plate is specified for a certain flow rate and therefore its accuracy is highest at this flow rate. Because of this limited accuracy to a single flow point and the improvements being made in electronic metering of liquids, new state-of-the-art techniques are being used more frequently. The latest trend is to install a solid state electromagnetic flow meter with a precise digital readout. The output of these devices has the advantage that they can be connected to facility DCS systems or other monitoring systems for continuous readout.

In dire circumstances, where such outlets are not specifically provided at the firewater pump or alternative flow meters are not installed, facility hydrants or hose reel outlets can be used. Flow measurement can be taken from the outlets of these devices in a similar manner to the 63.5 mm (2 ½ inch) outlets from a test manifold. Even portable clamp-on electromagnetic devices and ultrasonic flow meters are available.

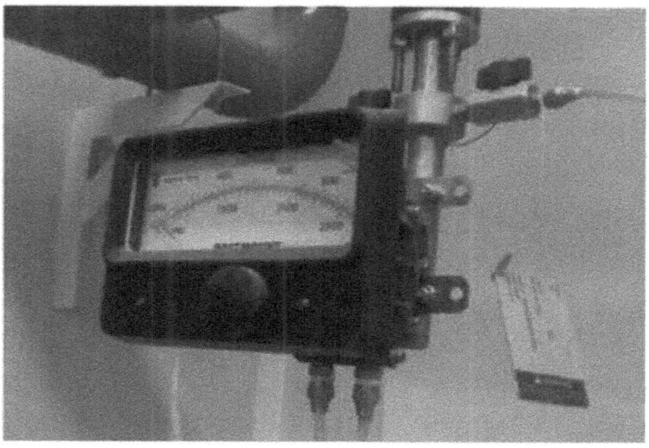

Figure 7-6 Firepump Flowmeter Device

Fixed flow meters should have a range of not less than 175 percent of the rated capacity of the firewater pump and preferably be listed. NFPA 20 lists specific sizes of meters to be used with rated pump capacities.

The ideal pump piping arrangement for flow testing is to design the piping so flow test water is recirculated directly back into the storage reservoir or supply source from which it has been extracted. This avoids undue setup requirements for testing and avoids wasteful use and spillage of water. The exact flow meter installation arrangement is usually guided by the requirement of the manufacture to ensure high accuracy of the device. In most cases, a straight flow of pipe is required for a number of pipe diameters from the inlet of the flow measuring device.

For offshore installations, a drainage test line is routed directly back close to the sea surface. This is because direct disposal under the structure may affect personnel who periodically work at the lower levels of the structure, or the water discharge may impact on the release of lifeboats in the vicinity in an emergency. Attempts have been made to route the disposal of firewater during testing to the annulus between the pump and column and outer protective casing. Although adequate space may be available for the water to be disposed of, the volume and velocity of the water being taken up cannot be compensated for quickly enough in the disposal cycle, and the water tends to collect in the annulus void and retard performance of the pump. This may also cause a slow heating of the water as it continually recirculates through the pump.

7.19. CHECK VALVES

Check valves (CV) or non-return valves (NRV) should be provided where there is a concern over damage to a pump from backpressure from the distribution system. This may be due to reverse rotation caused by unexpected driver failure or from the operation of another pump, which feeds into the same system. It is provided between each pump discharge flange and its discharge isolation valve. NFPA 20 requires all firewater pumps to be provided with a *listed* CV. A critical check during the construction and commissioning of firewater systems is the confirmation of the correct orientation of the check in the direction of water flow (i.e. that the CV is not installed backwards).

7.20. AIR RELEASE VALVE

An automatic air release valve (ARV) must be installed on the casing of a split case pump to release air from the casing of the pump when the pump starts and also admit air when the pump is shut down. Air in the intake of a pump causes cavitation and will lead to reduced performance of the pump and premature wear of the impeller. Considerable air will also be present in the column for a deepwell pump or a submerged offshore pump installation. The air release is required on pumps that start automatically and shall not be less than 12.7 mm ($^1/_2$ inch) in size. For vertical lineshaft pumps, the air release valve should be located at the highest point in the discharge line between the firewater pump and the discharge check valve. As per NFPA requirements, the ARV must be a listed or approved float-operated type device (see Figure 7-7). Some certifying authorities may also require the valve to contain non-combustible materials only (i.e. all metal), as sometimes the internal float mechanism is made of plastic.

Air entrainment is also a concern in the piping leading to horizontally driven pumps. For this reason, the suction piping is design to eliminate the formation of air pockets. Eccentric tapered reducers are provided in the intake piping and the piping should not be arranged above the pump so that air pockets can form and bring air into the impeller causing damage.

7.21. SUPERVISION OF ISOLATION VALVES

The closure of supply or discharge valves at firewater pumps completely prevents their capability to supply firewater. These valves are therefore considered critical in the

Figure 7-7 Horizontal Split Case Pump with Air Release Valve (mounted at top of pump).
Source: Photo courtesy of Patterson Pumps

Figure 7-8 Isolation Valve Fitted with Tamper Monitoring Position Device

reliability of the firewater pumping system to deliver water. Carseals or tamper switches can be provided to control inadvertent operation (i.e. closure) of the valves and are recommended by fire codes requirements and insurance underwriters. Activation of the tamper switch should cause an alarm indication at a remote monitoring location or control room (see Figure 7-8).

7.22. INLET SCREENS, STRAINERS AND FILTERS

Firewater pumps are not designed to handle foreign matter and can be easily damaged if foreign material is ingested into the impeller. Additionally, foreign matter can clog the pump and reduce its performance. Almost universally, firewater pumps are required to have an intake screen or strainer where they take suction from open bodies of water or from wells. These are commonly made of copper or brass to inhibit marine growth. Firewater pumps that receive water directly from public water supplies are not normally provided with strainers or filters, as the public water supply has been adequately treated and entrained materials have been removed (see Figure 7-9).

NFPA 20 requires a strainer to be provided at the suction manifold of vertical turbine firewater pumps. The strainer should have a free area that is at least four times (i.e. 400 percent) the area of the suction connections and will restrict the passage of a 12.7 mm ($^1/_2$ inch) or larger sphere. The interior passages of fire pumps are required to be not less than 12.7 mm ($^1/_2$ inch) in diameter, so no internal blockages will occur due to ingested objects. Basket shaped strainers are provided where pumps take suction from a pit. Where pumps are located in wells or seawater caissons, cone-shaped strainers are provided in order for the filter to fit into the borehole or caisson of the pump. Where

Figure 7-9 Strainers in Pump Inlet Lines

firewater pumps are supplied by a stream or lake and potentially damaging material may be present, a double or traveling screen is provided at the entrance to the pump water supply pit.

In some locations, an additional, secondary water filter is provided in the immediate discharge of the pump prior to its connection to the firewater distribution network. This is to ensure small fire suppression system orifices, such as sprinklers, hose reel nozzles, etc. do not become plugged as a result of particles smaller than 12.7 mm ($^1/_2$ inch), being distributed by the pump from the intake water, or scale and flakes from the corrosion of the pump or its suction or discharge pipes collecting in pockets in the system.

All screen strainers and filters need to be periodically inspected to ensure they have not become clogged or have deteriorated. One of the first items of investigation in the case of lower pump performance is the investigation of the condition of a pump intake strainer.

7.23. SUBMERGED PUMP INTAKE OPENINGS

Especially critical in fire pump installations that use open bodies of water is the activity of underwater diver operations in the proximity of the submerged fire pump suction bell or intake opening. Underwater diving operations routinely occur at the structural support (i.e. jacket) for industrial offshore installations for corrosion monitoring, modification inspections, etc. and in some locations for recreational purposes. A high water current occurs at the submerged intake of the pump when it is operating. This current poses a safety hazard to underwater divers who may be drawn into the submerged pump intake.

During operation of the ill-fated Piper Alpha offshore platform, it was common practice to switch the pump to a manual startup mode (thereby requiring an individual to visit the local fire pump control panel to start it up in an emergency) during diving operations. This was one of the contributing factors during the fire incident that entirely destroyed the platform. For reasons of diver safety, all firewater pumps that take suction in areas where underwater diver operations occur should be provided with suitable guards to preclude the necessity for placing a pump in a non-automatic start mode.

The International Association of Underwater Engineering Contractors has issued a Notice (AODC-055), describing the requirements for a large protective pump intake grid far enough away from the suction bell of the pump so that it will limit the water velocities to below that which would be harmful to divers. The object of the design is to prevent a diver, or in the case where diver umbilicals are used, to prevent the umbilicals being drawn into a pump water inlet and trapping divers or causing them injury. As it is common safety practice for two divers to work in close proximity to each other, consideration of any solution should take into account that a protective structure might be obstructed by both diver bodies and their equipment.

Additionally, the pattern of water flow at or near to a pump intake is influenced by various factors, however the maximum velocity usually occurs at the point directly in front of the pump intake. AODC recommends that a current of not more than 0.5 m/sec (1.5 ft/sec) be permitted for exposure to divers from the pump intake. Seamark or similar identification plates, that are readily visible, should also be fitted at the pump intakes to highlight their location for diving operations.

7.24. CAVITATION, NET POSITIVE SUCTION HEAD AND VORTICES

The most common problem associated with centrifugal pumps is believed to be cavitation. In all pumps, steps are taken to prevent cavitation and the formation of a vacuum that would reduce the flow and damage the structure of the pump. Avoidance of cavitation can be accomplished by the proper examination of pump suction conditions and calculation of the net positive suction head available (NPSHa). When a liquid is exposed to a pressure below its vapor pressure, it will begin to vaporize. Pockets of entrained air will begin to develop. As applied to pumps, it is commonly referred to as cavitation. It will initiate at a point in the piping or pump where pressure on the liquid approaches its vapor pressure.

When calculating NPSH, it is important to take the specific gravity of the fluid to be pumped into consideration. The level of atmospheric pressure exerted on the free surface (or enclosed reservoir) of the pumped liquid will also affect the dynamics of the installation. Table 7-5 provides data on atmospheric pressure versus altitude for water.

Vortices may also sometimes evolve in submerged pump suction bells in pump pits or tanks, causing cavitation, vibration or loss of efficiency. A vortex plate is normally fitted to these locations to reduce the possibility of vortices forming. All pumps taking suction from a stored water supply are required to have a vortex plate fitted at the entrance to the suction pipe.

Table 7-5 Atmospheric Pressure Versus Altitude

Altitude Meters (Ft)	Atmospheric Pressure psi	Equiv. Head in Ft. of H_2O	Boiling Point of H_2O °C (oF)
0 (0)	14.7	33.9	100 (212)
305 (1000)	14.2	32.8	99 (210)
610 (2500)	13.4	30.9	97 (207)
1219 (5000)	12.2	28.1	94 (202)

7.25. WATER HAMMER OR SURGE

Water hammer is the result of a rapid rise in pressure occurring in a closed piping system. It normally occurs as a result of sudden pump startup, stoppage (or failure) or from the change in speed of a pump or the sudden opening or closing of a valve, resulting in a change in water velocity in the system. In offshore facilities, a large amount of water accelerates up a long (empty) column and in some cases into an empty pump delivery pipework ring main. The problem may be exaggerated where there is air trapped in the main that acts as a spring or when a deluge valve opens. In the flow of fluids, the occurrence of water hammer, sometimes called surge, can cause damage to the distribution system unless adequate safeguards are provided.

An increase or dynamic change in pressure is produced as a result of the kinetic energy of the moving mass of liquid being transformed into pressure energy. This results in an excessive transient pressure rise (i.e. water hammer or surge). There can also be secondary surge problems associated with resonance, control valve interaction and the creation or collapse of vapor or gas pockets.

Technically, the rapid rise in pressure caused by a water hammer or surge effect is not necessarily a problem if it does not exceed the pipe rating. Piping codes allow for various design overpressures and higher pressure rated valves could be installed. However, they do not allow for the type of short duration pressures that might normally occur with continually repeated severe surge or water hammer conditions that progressively overstress the system. A pipe might be able to absorb some severe surge effects over a relatively short period of time, but the pipe could be weakened (due to repeated fatigue effects to the system) and may be expected to rupture sometime afterwards. Symptoms of the problem may be pipe movement, slamming of pump delivery check valves, and "hunting" of pumps and their drivers.

Elimination or reduction of water hammer effects can be obtained either by instituting controls on the sources of water hammer, release of the surge generation, or by accessories to absorb the impacts of water hammer without damage. Control on the start up and stopping of firewater pumps and the opening and closing of types of valves can be provided so neither of these operations will occur rapidly and result in the occurrence of water hammer. Additionally, surge control dump valves are available that release the pump start up or surge pressure and gradually close as the pump output reaches its rated level. Generally, the avoidance of surge is preferred over the absorption of surge in a system. Methods to absorb the effects of water hammer include surge chambers or vessels; however the system may still be suffering from the effects of surge, until the surge effect reaches these devices.

A Formal Interpretation (FI 83-10) of the NFPA 20 committee on firewater pump startup recommends the pump reach its rated speed during startup without delay. This is to avoid the possibility of a fire situation getting out of control during the short time in which a pump starts up, usually less than one minute. Therefore, the use of methods to avoid surge occurrence and which do not involve limitations on the pump startup sequence are gaining in popularity.

Computer modeling of firewater system transient pressure conditions can now be modeled and an analysis of surge conditions identified and generally easily remedied. The installation of a firewater pump into a distribution system should consider if the pump would produce surge conditions.

In the past, claims have been made from water hammer damage under "Extended Coverage Endorsements" as *explosion* damage under fire insurance policies/Insurance Policy Coverage. It was held that where water hammer was not specifically excluded from the extended coverage endorsement insurance policy, the subsequent insurance claim had to be honored (*L. L. Olds Seed Co. v Commercial Union Assurance Co.* 179 F. 2nd 472, 1950). Because insurance policies are a business contract, the exact interpretation of an explosion is decided in a court of law in these cases. It may conclude that a violent, outward bursting accompanied by noise would be considered to be an explosion. Insurance underwriters now try to clarify explosion coverage in extended coverage endorsements. They do this by additional wording to clarify that water hammer and the bursting of water pipes are not explosions within the intent or meaning of the provisions of the policy. If the insured desires this coverage, it is provided as a separate provision to the policy commensurate with the exposure that the insurance agent feels is justified.

7.26. PUMPING SYSTEM HYDRAULIC DESIGN

A variety of hydraulic software modeling programs are now on the market that can assist with the design and analysis of water supplies, pumping systems, distribution networks and pump surge or NPSH concerns. In a number of minutes, these programs can analyze a number of design alternatives for gravity, pump station and forced flow systems. Graphs of the pumping system, identification of possible pump cavitations or surge occurrences, and identification of specific point pressure and flow within a network can be easily provided. They offer evidence to the AHJ of the ability of a supply system to meet the demands of fire suppression systems and are normally included as part of the design documentation submitted for approval.

As with the use all software design programs, the input data is where an error or miscalculation may occur, as the formulas that perform the calculations have been proved over time. To overcome some of this concern, some programs now include standard pump, valve, pipe and fitting databases, along with standard pipe internal diameters for various piping classes and materials. One program also offers a feature to analyze up to 10 parallel pumps online.

The type of analysis performed should indicate the friction formulas used (Hazen-Williams versus Darcy Wiesbach), loop analysis methodology (Hardy Cross versus Newton-Raphson), limit of interactions, internal diameters and friction factors for piping and fittings.

7.27. VIBRATION LIMITATION

The Hydraulic Institute specifies acceptable limits for maximum permissible amplitude of displacement in any plane for clean liquid handling pumps. These limits (or curves) should be used as a general guide, keeping in mind that pumps which produce vibration amplitudes in excess of the indicated values should be examined for defects or possible corrections. The change in vibration amplitude over a period of time is often more important that the actual vibration amplitude itself. Vibration in excess of these values may be acceptable, that is, if there is no increase over a long period of time and if there is not any other indication of damage, such as increase in bearing clearance or noise level. The Hydraulic Institute recommend a pump be examined for defects when the vibration limits are exceeded.

Vibration measurements are made at the top motor bearing and bearing housing, respectively, for vertical and horizontal pumps. These measures should be accomplished at the time of factory acceptance of the pump and initial installation at the facility. Afterwards, periodic vibration monitoring results can be compared to the original installation to determine possible deterioration of the unit.

Diesel engine packages installed on offshore platforms may be subject to vibration impacts from surrounding equipment that can be harmful to the bearing of an engine if it is not operated for long periods, such as a firewater pump driver. They may also be needed if the engine is cited next to living quarters, where vibration from the pump driver would be a nuisance to the occupants. In such cases, the package (i.e. skid) is mounted on vibrational isolators.

7.27.1. TORSIONAL VIBRATION ANALYSIS

When vertical pumps are operated at various speeds, such as run-up to rated speed, the possibility of vibration increases sharply. Vertical pumps are very long and slender and their bases, discharge and supports have low natural frequencies. Therefore, natural vibrational frequencies, rotating unbalance, coupling misalignment, poor or worn bearings or hydraulic turbulence can exert alternating dynamic forces on the vertical structure. As the pump speed is changed, the frequencies of these dynamic forces will also change.

When the frequency of dynamic forces match the natural frequency of a pump component, resonant vibration will occur. The vibration produced in these cases is quite severe and can be visibly seen. The chance of these conditions occurring is much less if the unit is operated at a constant speed than if the speed is varied. Experience from offshore installations seems to verify that the starting and stopping of submerged pumps exerts the hardest stresses on the pump and power transmission system (i.e. vibration and up-thrust and down-thrust forces).

Computer modeling of torsional vibrational natural frequencies is now routinely accomplished for vertical pumps when requested. Vibration analysis starts with identification of when natural frequencies occur, usually identifying the first, second, third and fourth amplitudes for various operating speeds of the pump. The pump designer ensures that at the rated speed of the pump, these amplitudes do not readily occur or are sufficiently low so as not to be of concern.

Additionally NFPA 20 requires that a complete mass elastic system torsional analysis is undertaken for firewater pump systems that include a right angle gear drive to ensure there are no damaging stresses or critical speeds within 25 percent above or below the operating speed of the pump or drive.

7.28. BACKFLOW PREVENTION

Break tanks or double check valves are required when firewater pumps are prohibited from taking direct suction from public supplies. These direct connections are usually prohibited because of concerns for backflow of the fire protection system in the public main causing a health hazard, or because of hydraulic limitations of the public water supply system. A break tank provides an air gap between the public water supply and the private water service by providing a fill point above the normal water level. Double check (one way) valves improve the reliability of the prevention of backflow from the downstream to the upstream side during stagnant conditions. If break tanks are necessary, separate tanks should be provided for each pump.

7.29. AREA AND TASK LIGHTING

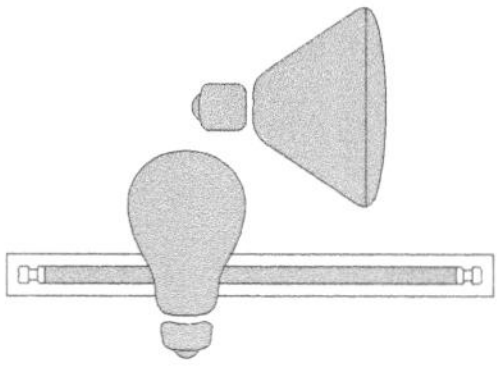

Pump areas should be provided with adequate lighting to operate the system, read instrumentation and manuals, perform service and maintenance on the unit, and safely evacuate the area in an emergency. Illumination levels should be as required by company standards or the minimum cited by the Illuminating Engineering Society (IES) Lighting Handbook.

Lighting requirements can be subdivided into general area lighting and lighting to perform specific tasks. General pumping areas are recommended to have illumination levels of 107.6 lumens/square meter (10 ft.-candles). Task lighting levels depend on the detail of work required to perform the task. Highly detailed work or difficult inspections can require illumination levels up to 2,152 lumens/square meter (200 ft.-candles). Emergency lighting units should also be provided in the pumphouse for evacuation in emergency conditions.

7.30. VENTILATION

A ventilation system should be provided for firewater pumps located in rooms or pump houses to ensure adequate air for the internal combustion engine-driven pump and to ensure adequate cooling for the driver and pump. Special consideration should be provided where acoustical enclosures are provided that may cause static air conditions to develop inside. When air supplies are provided through forced air systems, the power to the forced air system should be reliable during the fire incident. Where condensation may be a concern, an area room heater should be installed to prevent its formation.

7.31. FIRE SPRINKLER PROTECTION

Firewater pumps driven by diesel engines and contained within rooms or buildings are required to be protected by automatic sprinkler, water spray or foam water sprinklers. A design density of 0.17 L/sec-m^2 (0.25 gpm/ft^2) is specified for the fire pump risk area. Very recently, water mist protection systems have been used for the protection of firewater pump rooms with diesel drivers.

7.32. UTILITY SERVICES

Air, water and electrical power connections should be in or within a reasonable distance from the firewater pump installation. Routine maintenance activities for motors, pumps and engines require all of these services for efficient onsite activities. The connection points for these services should be adequately identified; especially the circuits for firewater pump services.

7.33. DRAINAGE

Drainage should be provided at a pump facility to expeditiously remove water from leakages or unexpected piping failures. The outlets of pressure safety valves should not be arbitrarily configured, but be routed to the nearest drain or external area for water disposal. The floor should be sloped away from the pumping area to a nearby drain that is not susceptible to freezing. The drain system should be provided with suitable seals or traps so that vapors from process area sewers are not released into the firewater pumping area.

The pump room itself should be located on high ground which is not subject to flooding. Pumphouses should be provided with roofs that slope, so that rainwater does not collect on the roof and possibly seep onto pumping equipment, causing damage.

7.34. OUTSIDE INSTALLATIONS

Firewater pumps located outside are generally only provided in temperate climates where they will not be vulnerable to fire and explosion incidents. Protection from excessive sun and or rainfall is usually afforded in the form of an overhead canopy for these installations. Precautions against animal and insect nesting in the equipment must also be provided.

8 MATERIALS OF CONSTRUCTION

8.1. INTRODUCTION

The metallurgy selected for the construction of a firewater pump is primarily dependent on the properties of the water source to be used, economics and the life of the installation. There are no national codes or regulations that state pumps must be made from certain materials, although a company may have its own specifications and guidelines. For fresh water sources (i.e. public water mains, fresh water lakes, etc.), cast iron is normally adequate although bronze internals may be optional. Brackish or seawater utilization will require the use of highly corrosion-resistant materials and possibly coatings. Typically, specified materials include alloy bronze, Monel, Ni-resistant, stainless or duplex stainless steels combined with a corrosion-resistive paint or coating. However, having selection of one of these does not guarantee a highly corrosion-resistant pump as each of these materials has its own limitations.

The most important environmental conditions to consider when selecting pump materials are the water velocity (i.e. stagnant water concerns), galvanic effects (dissimilar metals), aeration and marine environments.

In seawater pumps, the most critical component in terms of pump operation is the rotating element, since minor amounts of corrosion on impeller edges can effect efficiency and result in pumps running out of balance. In general, the pump bowl or casing can stand significant amounts of corrosion before operational problems are observed.

Alternatively, the second most vulnerable area for corrosion in offshore firewater systems exists in the splash zone. The splash zone is the area where the water surface is continually active, the waves become highly aerated and much water spray is generated. This produces more corrosion and erosive effects than in a submerged location. Splash zone corrosion rates vary according to location, increasing with the amount of wave action and seawater aeration. As the splash zone is only intermittently immersed in water, the use of a protective coating is required on riser pipework (see Figure 8-1)

8.2. DURABILITY

The major features that will lead to a long pump life are identified as follows:

- Pumping "neutral" liquids at low temperatures.
- Elimination of abrasive particles in the pumped fluid.
- Continuous operation at or near the maximum efficiency capacity of the pump.
- An adequate margin of Net Positive Suction Head (NPSH) available over NPSH required.

Fire Fighting Pumping Systems at Industrial Facilities. DOI: 10.1016/B978-1-4377-4471-2.00008-6

Figure 8-1 Offshore Platform Detail of Splash Zone

A system that has the pump continuously cycling, using hot corrosive liquid containing abrasive particles with a low NPSH available, will obviously undergo a lot of stress and wear compared to a pump with the opposite features. Therefore, all the materials chosen for the pump construction, i.e. impeller, casing, bearings and other components, should be chosen based on these influences and how long the facility is to be in existence.

Components that need high strength and dimensional stability, such as internal combustion engines and gear drives, may require heat-treated and strain-relieved alloys.

8.3. CORROSION CONSIDERATIONS

Corrosion is broadly defined as the deterioration of a material due to a reaction with its environment, where deterioration implies a change in the structural properties.

Dissolved oxygen and saltwater from seawater is perhaps the greatest factor in the corrosion of steel surfaces in contact with oceans and seas. Common coated cast iron pumps in seawater service conservatively have a two to three year life, while pumps made of nickel-aluminum-bronze are expected to have a seven to 10 year lifespan. Therefore, careful consideration of pump materials should be undertaken to avoid premature failure of the system.

Locations where water and air come into frequent contact will have more corrosion activity in general than other locations. Common locations where this occurs are at splash zones, rotating parts submerged in water, and air leakage points into the system. Therefore, most corrosion control effects are directed to counter these effects. Additionally, pumps operating continuously in a given medium often exhibit a lower rate of corrosion than pumps operating intermittently. During stoppage, as in the case of firewater pumps, air may more readily accumulate inside a pump, accelerating corrosion. Where corrosion is a concern because of stagnant water conditions, on some occasions it may be useful to maintain a small flow through the system from the discharge of the jockey pump during periods when the firewater pumps are shut down.

Fretting corrosion may occur from the vibration of bearings pressing against the rotating components. Diesel engine packages installed on offshore platforms may be subject to vibration impacts from surrounding equipment. This is harmful to the bearings of an engine that is not operated for long periods, such as a firewater pump driver. In such cases, the package is mounted on vibrational isolators.

8.4. CATHODIC PROTECTION

Seawater pump risers that are submerged may be provided with sacrificial aluminum alloy anodes and impressed currents to inhibit corrosion. These types of systems are commonly employed on large offshore steel platforms (primarily for oil and gas production). Galvanic action promoted by the presence of seawater will decompose the anodes in preference to the steel of the riser, thereby protecting it.

Risers are particularly vulnerable at the seawater level where the constant wave action creates an oxygen-enriched water zone around the riser. Periodic underwater inspection and replacement of the anodes is necessary to ensure that riser corrosion does not occur. The inspection consists of evaluation of the anode condition and measurement of the potential difference generated. In addition, maintenance of the impressed current is required.

8.5. COATINGS

Epoxy coated steel column pipes from submerged vertical turbine pumps has proven to be a disaster. Whenever a break in the coating occurs, corrosion is concentrated at the break point and the rate of corrosion is greater than if the steel had not been coated in the first place. Plastic coating of impellers and bowls has likewise not proven to be successful.

8.6. FIBERGLASS MATERIALS

Fiberglass materials (e.g. fiberglass composites, Reinforced Thermosetting Resin (RTR), etc.) have recently been used for firewater systems, including the intake and supply piping of firewater pumps, especially for offshore installations (with appropriate fire rating certificates). They have been found to withstand the harsh marine environmental effects and offer superior corrosion resistance. The protective caisson of the pump column has similarly been supplied in fiberglass materials on occasion.

8.7. FRESH WATER CONCERNS

In fresh water supplies, the primary concerns are alkalinity and abrasiveness.

The alkalinity or acidity of a water source can be assessed by its pH level. In general, a pH above 8.5 or below 6.0 precludes the use of a standard bronze-fitted pump (i.e. cast iron casing, steel shaft, bronze impeller, wearing rings and shaft sleeve). Groundwater is often associated with a high pH reading, consequentially all iron or stainless steel fitted pumps are used.

Abrasiveness is a result of suspended matter or sand in the water supply. Due to the fact that these particulates are small, they will pass through the inlet strainer or the pump. This may require the selection of a stainless steel or nickel-cast iron casing, cast iron nickel cast iron or chrome steel impellers and stainless steel, phosphor bronze or Monel wearing rings, shafts, sleeves and packing glands. The fitting of a filter is sometimes employed to collect the suspended matter in the water supply before it reaches the firewater pump intake; however these can be easily plugged and should be carefully monitored if employed.

8.8. COMMON PUMP MATERIALS

With the wide range of materials that are currently available, virtually all corrosion problems can be technically resolved. It is economic considerations that must also be considered when material selections are being made. Most companies usually have

Table 8.1 Common Firewater Pump Materials

Water Service	Casing	Impeller	Shaft
Freshwater: Option #1	Cast Iron	Bronze	Stainless Steel
Freshwater: Option #2	Special Grade Iron	Bronze	Stainless Steel
Seawater: Option #1	Ni-Resist	Stainless Steel	Duplex Stainless Steel
Seawater: Option #2	Duplex Stainless Steel	Duplex Stainless Steel	Duplex Stainless Steel
Seawater: Seawater #3	Nickel Al. Bronze	Nickel Al. Bronze	Duplex Stainless Steel

Notes:
1. Cast iron materials should only be used when non-aggressive fresh water sources are utilized.
2. Duplex stainless steels types are normally used that have 25 percent Cr.
3. Hypochlorite solutions should be used with bronze materials.
4. For seawater duties Monel K-500 may be substituted for Duplex Stainless Steel for shafting applications.
5. Materials for all pump components should be compatible with each other and the overall system.

Materials that have been used for firewater pump construction for seawater corrosion resistance have ranged from common 316 Stainless Steel and Nickel-Aluminum-Bronze to 14462 Zeron 25, 6MO Steels and Super Duplex Zeron 100.

their own preferences for the types of materials they prefer for construction of their firewater pumps. These are usually based on their own experiences within their company and evaluation of international references and standards. Service life and economics are also factors that are considered. Material delivery times may also be important where short project installation schedules are involved. Table 8-1 lists commonly encountered firewater pump materials specified by industry in general.

9 PUMP DRIVERS AND POWER TRANSMISSION

9.1. INTRODUCTION

All pumps require some means of power to rotate their impellers in order to impart momentum to the pumped liquid. The primary power used is customarily either a diesel engine or an electrical motor. These devices can be configured in various arrangements and designs. By far the most common are electrical motors in highly industrialized areas and diesel engines in remote or undeveloped areas and as backups to the electric drivers. Steam drives are used where steam supplies are conveniently available and considered reliable.

All drivers need a mechanism to transmit their produced power to the pump impeller. Again, various methods are available to connect these devices. Couplings, lineshafts or hydraulic fluids are used to transmit power from the driver to the pump.

9.2. ELECTRIC MOTORS

The preferred driver for most fire pumps at industrial facilities, when there is a reliable and non-vulnerable power grid available, is an electrical motor that receives energy from two different power sources (i.e. generation stations) (see Figure 9-1). Where the available electrical power grid is unreliable, or from a single source, fire pumps powered by high horsepower diesel engines should be provided. Nowadays the power generation plant and power grids of industrial countries are considered highly reliable and motors are of high quality and reliable, so the need for an independent prime mover is not as severe as it may have been several decades ago. By examining the number and duration of outages and the arrangement and redundancy of the power grid, a judgment as to the reliability of a power supply can be made. Further examination of onsite and offsite fire exposures to power transmission lines that supply firewater pumps also needs to be undertaken as part of this examination.

Whenever electrical motors are selected to drive firewater pumps, the power supply should be provided from dual power lines. Preferably, multiple electrical pumps should be fed from different feeders which would not be simultaneously impacted from any major incident at the facility. The firewater pump's electric motor should be connected to the plant emergency power supply if it is available and of sufficient capacity.

From the point of view of low maintenance, easy startup and dependable operation, electric motors are preferable to other drivers. The maintenance, failure points, fuel inventories, instrumentation and controls needed for an internal combustion engine

Fire Fighting Pumping Systems at Industrial Facilities. DOI: 10.1016/B978-1-4377-4471-2.00009-8

Figure 9-1 Typical Electrical Centrifugal Fire Pump

versus an electrical motor all demonstrate that it is not a cost effective option as compared to an electric motor.

Industrial motors are generally three-phase, squirrel cage, induction motors. The induction motor operates by an induced current in the rotor, which causes a force to move it in the same direction as the impressed electrical field. In application, the rotor is unable to maintain speed with the field and some slip will occur; this increases the amount of torque developed. A two-pole motor driving a centrifugal pump, for example, may turn at 3,500 revolutions per minute (rpm) instead of 3,600 rpm.

Two arrangements of electrical motors are available for firewater pumps, depending on the source of firewater supply and other factors; namely direct electric motor connection to a firewater pump from the rotating shaft or an electro-submersible pump connected by a cable to a dedicated diesel driven generator set. Direct electrical motors are becoming the most common firewater pump drivers in industrial countries where the commercial power grid or facility onsite power generation supplies are considered very reliable.

Electro-submersible pumps are used for well supply sources or in offshore installations. In these cases a diesel power generator is provided topside and an electrical cable is routed to a submerged electro-submersible pump located in the ocean or well. The submerged pump and motor is suspended at the bottom of a column pipe in a caisson. The pump is driven by a submerged motor located below the pump bowls. All thrusts from the pumps are accommodated by the motor.

Due to the increased reliability of submerged electrical pumps in the last decade, the popularity of this style of firewater pump has increased in offshore use, because of the installation advantages it offers. Electro-submersibles require good sealing of the motors and cable connections. The cable must be protected against environmental attack by the sea, and mechanical damage and fire exposures at the topside facility.

Submersible pump motors (either electric or hydraulic) require double mechanical seals.

Alternating current (AC) motors are more efficient and cost effective than direct current (DC) motors and are almost universally selected to drive all motor driven firewater pumps. Generally, only small pumps are designed to use DC supplies, of 12 or 24 volts, in order to utilize battery power sources or where it is the only power source available. Because of their small capacity, they are not well suited to supply firewater needs.

Electrical motors operate successfully where the voltage variation does not exceed 10 percent above or below normal, or where the frequency variation does not exceed five percent above or below normal. The sum of the voltage and frequency variation should not exceed 10 percent.

Power cables to firewater pump motors should preferably be buried or routed away or protected from designated fire sources, potential explosion areas or similar exposures. The feeder circuit should be independent so that plant power can be isolated or shut off with interruption to the firewater pump. All circuits that supply or support firewater pump services should be highlighted and adequately marked.

9.2.1. NEMA CLASSIFICATION

Electrical motors must be built to operate under the conditions they may expect to be exposed to, such as fumes, splashing liquids, and airborne solids. The National Electrical Manufacturers Association (NEMA) in the United States has therefore provided a classification system to define the type of exposure a pump may be subjected to. These are highlighted in Table 9-1.

9.2.2. SPLASH SHIELD OR PARTITIONS

Where water from a pump seal can splash water onto an indoor electrical motor which is not rated for outdoor exposures (i.e. rain), a splash shield or partition has been commonly provided in the past. The splash shield is a circular steel plate provided to deflect splashing of water away from the motor housing, to prevent it from shorting out. The National Fire Protection Association (NFPA) 20 recommends that motors that may be or are subject to the splashing effects of water should be completely enclosed.

9.3. GASOLINE ENGINES

Internal combustion engines, powered by flammable fuels such as gasoline, natural gas and liquefied petroleum gas (LPG), are no longer recommended. Although provided and allowed in the past, since 1974 they are no longer recommended by many local authorities or recognized by NFPA 20 as a driver for firewater pumps. This is due to the increased fire hazard they pose inside buildings through their use of a flammable fuel and the increased reliability that is gained by the use of a diesel engine or electrical motor in their place. Gasoline engines are also not commercially produced in very high

Table 9-1 NEMA Exposure Classifications

Type	Description
1	General Purpose
1A	General Purpose (Semi-Dust Tight)
1B	General Purpose (Flush Type)
2	Drip Proof (indoors)
3	Dust and Rain Tight and Sleet (ice) Resistant (outdoor)
3R	Rain Proof, Sleet (Ice) Resistant (outdoor)
3S	Dust and Rain Tight and Sleet (ice) Proof (outdoor)
3X	Dust Tight, Rain Tight, and Sleet (Ice) Proof – Outdoor, Corrosion Resistant
3RX	Rain Tight, and Sleet (Ice) Proof – Outdoor, Corrosion Resistant
3SX	Dust Tight, Rain Tight, Ice Resistant, Corrosion Resistant
4	Water and Dust Tight
4X	Water and Dust Tight and Corrosion Resistant
5	Dust Tight and Water Tight
6	Submersible
6P	Prolonged Submersible
7	A, B, C, or D Hazard Groups Class I (air break)
8	A, B, C, or D Hazard Groups Class I (oil-immersed)
9	Hazard Group E or G, Class II
10	Mine Safety and Health Administration Explosion Proof
11	Acid and Fume Resistant (oil immersed)
12	Industrial Use
12K	Industrial Use, with Knockouts
13	Oil and Dust Tight (indoor)

power levels, i.e. above 224 kilowatts (300 hp), as is required for some high capacity firewater pump driver installations.

9.4. DIESEL ENGINES

Loss incidents have shown diesel driven firewater pumps to be the most reliable in severe or catastrophic incidents (see Figure 9-2). Electric and steam drivers are more likely to be affected in a major incident since they rely on some portion of the facility's infrastructure, which more than likely will be impacted during the incident. Also, since power failures usually cannot be completely disregarded, backup or reserve capacity firewater pumps, meeting the full high risk water flow demands, driven by a diesel driver, should be provided at all major high hazard complexes or critical facilities. Petroleum and petrochemical operations in third world locations are generally dependent on their own power generation facilities, so self-contained diesel-driven units are selected to decrease sizing and costs associated with production power generators. Installation of diesel engines are generally made to the requirements of NFPA 37,

Figure 9-2 Typical Diesel Driven Fire Pump

Standard for the Installation and Use of Stationary Combustible Engines and Gas Turbines and NFPA 31, *Standard for the Installation of Oil Burning Equipment.*

Diesel engines utilized for offshore environments should be specified with a marine option. This feature improves the corrosion resistance of components used.

9.4.1. ENGINE GAGE PANEL

A local panel should be provided at the diesel engine, conveniently sited for easy observation, to provide status indicators and instruments on the operation of the unit (see Figure 9-3).

These instruments typically include the following:

- Tachometer w/hour meter (non-resettable).
- Lube oil pressure gage.
- Lube oil temperature gage.
- Cooling water temperature gage.
- Charge air manifold pressure gage (optional).

The unit chosen for the engine gages should be those commonly in use at the facility (i.e. Metric or English).

9.4.2. DIESEL ENGINE FUEL SUPPLIES

The diesel engine has to be provided with a fuel tank that can meet the demand of operating the unit during an emergency without refilling. Most references cite the

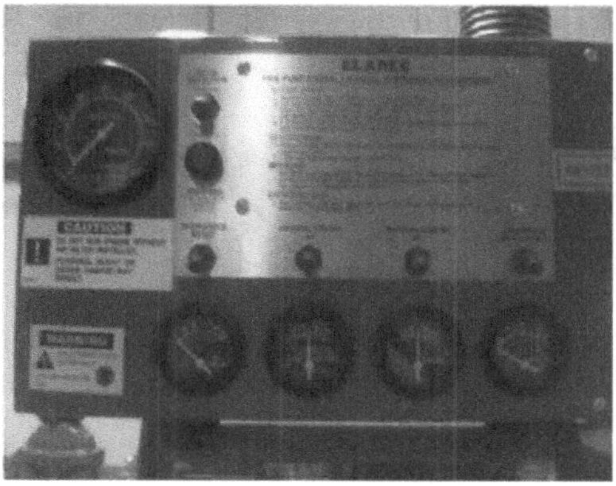

Figure 9-3 Typical Diesel Engine Instrument Panel

provision of fuel in the amount of 5.07 liters/kilowatt (one gal./hp) rating of the engine for eight hours (e.g. NFPA). A prudent examination should be made to determine the exact amount of fuel storage and refilling requirements.

The onsite fuel supply at the diesel engine should consider the following requirements:

- Maximum duration of the largest fire risk plus a safety factor.
- The minimum required by national or local codes.
- The amount recommended or stipulated by insurance underwriters.
- The amount required by in-company requirements.
- The time required for the fire pump to completely draw down stored firewater supplies that cannot be immediately replenished.
- A minimum of at least 5.07 liters/kilowatt (one gal. per horsepower) rating of the firewater pump driver (equivalent to one pt. per horsepower for eight hours).

NFPA 20 also requires that five percent of the tank be reserved for a sump, which is not to be used by the driver and an additional five percent should be provided for expansion. Therefore, all supplies must calculate an additional 10 percent capacity. Typically, an eight-hour duration is cited as the minimum, with supplies up to 18 hours mentioned in some installation code requirements and some with as little as 30 minutes (see Table 9-2). Some facilities have been provided fuel for up to 24 hours of maximum continuous operation of the engine.

The high levels of fuel supplies for ships and offshore platforms (18 and 12 hours) assumes that refilling capability may not be readily available immediately after an incident, that some incidents may last for a considerable length of time and essentially unlimited water supplies are available (i.e. the ocean).

Most diesel fuel supplies are provided adjacent to or in the fire pump room or location, if this is allowed by local regulations. A splash or leak collection pan is usually placed underneath the tank to retain any fuel spilled or leaked. Fuel is normally

Table 9-2 Pump Driver Fuel Duration Requirements

Standard	Fuel Duration Requirements
API RP 14G	30 minutes
FM Handbook (2nd Edition)	8 hours
Lloyd's Register	18 hours
NFPA Handbook (20th Edition)	8 hours
NFPA 20 (2010) (direct diesel driver)	1 pt/hp + 5% for Sump + 5% for expansion for 8 hours
NFPA 20 (diesel generator for electric pump)	8 hours
SI 611 (UK)	12 hours
SOLAS (Ships)	18 hours
UK DOT (Ships)	12 hours

fed by gravity to the engine, therefore most fuel tanks are provided on supports that are higher than the engine consumption point, within the pump room.

Where multiple, diesel driven, firewater pumps are provided, each should be provided with its own dedicated fuel storage tank. All piping for supply and return lines to the engine should be adequately protected against physical damage and comply with the manufacturer's recommendations. Where flexible connections are provided, they should not be rubber tubes but braided wire hoses. Fuel filters should be easily accessible for maintenance activities.

Gage glasses are prohibited on fuel tanks for firewater pumps by NFPA 20, due to the fact that the glass may be broken and all the fuel in the tank may drain out. A low fuel level alarm can be provided on the fire pump controller panel to warn of a low fuel level in the storage tank.

9.4.3. FUEL REFILLING ASPECTS

In instances where the fuel storage cannot be replenished in a reasonable amount of time, a reserve supply of fuel should be provided. A mechanism to transfer fuel from the reserve storage to the immediate diesel supply tank to the engine should be available. Some larger facilities have header for the distribution of diesel fuel to all the prime movers at the installation.

The fuel tank for the fire pump can be fitted with an automatic fill valve (float valve) to fill the tank once a pre-set low level is reached. Complete reliance on automatic refilling of the fuel tank by a float valve should not be undertaken, since the float valve mechanism becomes stuck in a high position (because of corrosion or other factors) and the tank will not be automatically refilled. Operator surveillance of local fuel levels should be conducted as backup. Drain and vent capability should also be incorporated on the storage tank.

9.4.4. FUEL CONTAMINATION

Diesel fuel can be contaminated from several sources. Microbial contamination can occur, which is caused by micro-organisms from foreign matter in the fuel with the presence of moisture and heat. They can cause fuel line plugging and corrosion. Fuel supplies vary by season and summer grades may not be as effective for winter seasons if the fuel has not been consumed by then. Partially filled fuel tanks can allow condensation and corrosion products to form. Water or corrosion particles can accumulate in the system as a result.

9.4.5. ENGINE STARTING SYSTEMS

Engine starting systems may be electric, hydraulic or pneumatic, provided they are reliable. To improve reliability, all of these systems should use stored energy to provide an immediate self-contained ability to force-start the engine with backup reserve supplies. Electrical starters use batteries, hydraulic systems use reservoirs and pneumatic systems utilize compressed gas reservoirs or high-pressure cylinders. Direct connection to facility utility systems providing these services should not be relied on because these may be impacted or unavailable during an emergency. The capacity of these stored energy sources should be sufficient to provide six starts of the engine within 30 minutes, with at least two starts within the first 10 minutes. Overall, battery supplies are preferred and commonly provided. This is primarily due to ease of installation and reliability. They may be supplemented by hydraulic or pneumatic sources (or replaced when there is a concern of an ignition source from battery usage during potential massive combustible vapor cloud releases) for reliability improvement of starting systems.

- Electric: Electric starting systems utilize storage batteries equipped with a trickle charger to maintain adequate power. Meters should be provided to indicate the charge provided from the batteries. NFPA 20 recommends the provision of two storage battery units for diesel engines. These are to be charged from two different power sources (i.e. one from the driver and one from the facility power).
- Pneumatic: High-pressure nitrogen or air storage cylinders, regulated to the required pressure via an intermediate expansion drum, are provided for pneumatic systems. A pressure gage should be provided to indicate the condition of the system. Where compressed air is used, the system is normally connected to the plant instrument air system to maintain system pressure.
- Hydraulic: Hydraulic systems utilize reservoirs pressurized by manual or automatic pumps. A pressure gage should be provided to indicate the condition of the system.

9.4.6. STARTING BATTERIES

Automotive storage batteries are provided to all diesel engines for independent starting power. As a minimum, two sets of batteries are provided to ensure reliability. One primary and one backup set. Should one set fail, the firewater pump controller cycles to the other battery set for a second start attempt. These batteries are usually placed on either side of the pump skid in designated battery compartments. Voltage and ampere gages are provided on the controller to indicate battery charging condition.

The starting ampere-hour and cold cranking amperes for engine starting batteries should be specified by the manufacturer of the engine. Either low maintenance nickel-cadmium alkaline batteries or lead acid types may be selected. Nickel-cadmium alkaline batteries are advertized to last for up to 20 years, offsetting their initial high cost against the relatively inexpensive lead acid types that usually last from three to five years. Lead acid batteries are more susceptible to overcharging and require frequent water replacement for their cells. They also may develop corrosion, and hydrogen gas is generated during charging.

Maintenance-free batteries are not recommended because their condition cannot be easily assessed without subjecting them to a load, and they cannot be cycle charged without a permanent loss of capacity.

9.4.7. ENGINE COOLING SYSTEM

Diesel engines must be provided with an adequate cooling system in order to perform reliably and satisfactory. A self-contained cooling system is normally provided, consisting of a heat exchanger, circulating pump and circulation piping. Common practice is to use a portion of the produced pump firewater to cool the engine. If this design is applied, the portion used for engine cooling should not be included for firewater delivered by the pump for fire-fighting purposes. Other pump installations provide a cooling radiator and use a self-contained cooling system for the engine. Self-contained engine cooling water may also be circulated into a separate heat exchanger which is then cooled with the produced firewater.

It should be noted that the horsepower requirement is greater for a radiator with a fan cooling system than with a water-cooled heat exchanger system. Water quality is also important when considering the use of a cooling system. Heat exchangers are prone to plugging where poor water quality is encountered. Self-contained cooling systems may use an ethylene glycol additive to achieve higher jacket temperatures and for freeze protection for the cooling water.

Engine cooling systems are commonly fitted with an engine block heater to maintain the coolant near its operating temperature and allow the warmed coolant to circulate through the engine. By maintaining the coolant in the engine at a temperate of about 49 °C (120 °F), it allows a quick start of the engine, reduces engine wear, and maintains the engine near its operating temperature of 60 °C (140 °F).

9.4.8. ENGINE EXHAUST SYSTEM

The purpose of the exhaust system is to remove hot burnt gases from the combustion chambers of the diesel engine to a remote disposal point, for dispersion in the atmosphere without harm to personnel or equipment. It is composed of an exhaust manifold, in some cases an expansion and vibration bellow, muffler and exhaust piping.

Diesel engines require a dedicated and independent exhaust system (see Figure 9-4). The exhaust system outlet should be located outside the pump room or house. The outlet should also be arranged to be as high above ground level as is practical, to aid in exhaust gas dispersion. This prevents exhaust fumes from affecting

Figure 9-4 Diesel Fire Pump Exhaust System

personnel, re-circulating back to the driver air intake, and other heating, ventilation and air conditioning (HVAC) or prime mover air intakes. In addition, it avoids the ignition of nearby combustible materials, or adversely affecting the operation of other equipment in the vicinity of the fire pump. Cases have been recorded where the exhaust of an engine has activated the smoke detectors in areas adjacent to the fire pump room because the prevailing wind directed the exhaust to these areas. The outlet point should also be arranged to exclude the collection of rainwater.

Exhaust systems that may come into daily contact with personnel should be insulated to prevent burns. Maintenance activity instructions should highlight the possible exposed hot surfaces of the exhaust systems and necessary precautions. Common practice is to provide exhaust system insulation or other suitable guards for personnel protection up to a level of two meters (6 ft). Personnel should not be exposed to surfaces temperatures greater than 65 °C (149 °F) that are within easy reach.

Those systems that are located within the areas of possible combustible vapor release or that have outlets directed towards combustible materials, should also have the piping insulated and the exhaust gases cooled to prevent them acting as an ignition source (i.e. from hot surface contact). Many incidents have been identified in the Gulf of Mexico where a fire ignition source has been the hot surface of an exhaust system. Chapter 12 provides further details of electrical area fire pump installation requirements.

The amount of allowable engine exhaust backpressure is also specified by the engine manufacturer for each specific engine. The exhaust gas backpressure requirements limit the length of exhaust piping or muffler sizes and arrangements that can be provided to a fire pump installation. Excessive backpressure will adversely act against an engine piston and reduce the amount of power available from the engine.

Exhaust system piping is usually composed of carbon steel. This piping may be expected to last up to 25 years in ideal conditions. Where there is extended use, or harsh or corrosive conditions exist (such as in a marine environment), the exhaust system piping may have a much shorter life span. Stainless steel piping is sometimes specified in these circumstances.

The primary purpose of the muffler is to lower the sound of the internal combustion engine exhaust noise to an acceptable level. The noise level limitations should be specified at the time of engine specification approval. Chapter 14 discusses further aspects of noise concerns.

Large or extensive exhaust systems may be prone to the collection of fluids through condensation of exhaust gases or entrained liquids (oils) in the exhaust. These fluids will collect or settle in the low point of the exhaust system, usually the muffler. Large mufflers are provided with a built-in sump to collect these fluids (as recommended by NFPA 37). A drain line is typically connected to the exhaust muffler sump where the drain port of the sump is inaccessible, to allow the periodic removal of these fluids.

Under certain conditions, the exhaust system may collect combustible vapors (unburned from the combustion chambers of the engine) and a "backfire" may occur. This backfire may allow nearby combustible materials to ignite if not controlled. The system should be able to withstand the explosion of unburnt gases in the exhaust system piping. Furthermore, if the muffler can be considered a source of sparks, then suitable provisions to arrest these sparks should be provided.

A suitable flexible connection from the engine exhaust manifold to the exhaust piping (i.e. an expansion bellows) to allow for movement of the driver during startup and shutdown, vibration and thermal strain is needed. This prevents the failure of the exhaust system causing a release of vapors within the pump area.

9.5. AIR SUPPLIES AND VENTILATION

Suitable quantities of fresh air are required for the combustion process of the engine and for cooling. These air supplies should be arranged so that they do not draw in the exhaust of the engine. Preferably, they should be provided from an elevated, upwind location to prevent the ingestion of vapors from accidental releases. Dust and sand filters should be provided where conditions warrant, to prevent damage to internal engine parts. If the engine is located within a dedicated building for the fire pump, intake of air from the building interior is acceptable.

The maximum ambient temperature at the fire pump location should be at or below 49 °C (120 °F).

9.6. INSTRUMENT PANEL

An instrument panel should be provided to all engines to indicate basic operation of the unit. As a minimum, the panel should be provided with a tachometer, water temperature and oil pressure gage in order to determine the basic running condition of the engine. The tachometer should also indicate total hours for the engine or a separate hour meter should be provided to account for maintenance periods and accountability of failure conditions.

More elaborate panels on complex or expensive engines may incorporate some or all of the following gages:

- Fuel pressure.
- Lube oil pressure.
- Lube oil temperature.
- Air inlet manifold temperature.
- Exhaust pyrometer.
- Fuel filter differential pressure.
- Lube oil filter differential pressure.
- Air filter differential pressure.
- Coolant water level.

The panel should be located where it is easily observable to the personnel that attend the operation of the engine. Gages should be marked in the units that are adopted for the facility (i.e. Metric or English).

9.7. STEAM TURBINE

Refineries and chemical plants make more steam than any other industry in the world. Steam is primarily used in these industries for process heat exchangers, power to drive pumps and compressors, electrical power generation and for purging, cleaning and inerting. In fact, steam supplies power to about 25 percent of all the pumps and compressors that move liquids and gases around refineries and chemical plants.

Where surplus steam supplies are available and reliable, steam-driven fire pumps can be effectively used. In industry, steam driven firewater pumps may be found were steam is heavily used in the operation of the facility. These pumps are often located at the source of the steam, i.e., the power plant or the boiler station.

Steam turbines for industrial uses are designed to meet the requirements of API Standard 611, *General Purpose Steam Turbines for Refinery Service*. More than one boiler capable of meeting the firewater pump demand should be provided. With boilers that have steam pressures of 840 kPa (120 lbs) or less, the steam turbine should be capable of driving the pump at rated capacity with steam pressures as low as 560 kPa (80 lbs). With boiler pressures above 840 kPa (120 lbs), a pressure of 70 percent is usually taken instead of 560 kPa (80 lbs). The steam turbine selected for the firewater pump support should not have a rated speed more than 3,600 rpm, since listed firewater pumps are not rated for speeds above this point.

Since a major power failure can cause too many turbine drivers to start up, thus depleting steam supply to the point that it is insufficient for the firewater pump driver and essential process use, a control system should be available to establish priority use.

The onshore options from positive suction supply are as follows:

- Horizontal Split Case Pump with Diesel Engine.
- Horizontal Split Case Pump with Electric Motor.
- Horizontal Split Case Pump with Steam Turbine Driver.
- Vertical Split Case Pump with Diesel Engine.
- Vertical Split Case Pump with Electric Motor.
- Vertical Split Case Pump with Steam Turbine Driver.

The onshore options from lift supply are as follows:

- Lineshaft Pump with Diesel Engine.
- Lineshaft Pump with Electrical Motor and Dedicated Generator.
- Lineshaft Pump with Steam Turbine Driver.
- Electro-Submersible Pump Unit with Dedicated Topside Power Generator.
- Submerged Hydraulically Driven Pump Unit with Dedicated Topside Hydraulic Driver.

The fixed offshore structure options are as follows:

- Lineshaft Pump with Diesel Engine.
- Lineshaft Pump with Electrical Motor and Dedicated Generator.
- Electro-Submersible Pump Unit with Dedicated Topside Power Generator.
- Submerged Hydraulically Driven Pump Unit with Dedicated Topside Hydraulic Driver.

9.8. POWER TRANSMISSION OPTIONS

Power transmission to a pump can be by direct mechanical couplings, right angle gear drives, or by indirect hydraulic means. Each of these mechanisms has its own advantages and suitability for a particular pump installation.

9.8.1. DRIVER PUMP COUPLING

If a pump and driver are rated to the same rated speed they may be directly coupled, otherwise a suitable geared coupling must be provided. Impellers which are mounted directly on a driver shaft extension are considered "close-coupled" units. Otherwise, rigid or more commonly flexible couplings are used.

All exposed moving portions of the firewater pumping system should be suitably guarded to prevent injury to personnel. Drivers to pump couplings or gear drives are normally provided with a shield or guard that can be easily removed for access to the unit during maintenance and repair.

9.8.2. VERTICAL TURBINE PUMP DISCHARGE HEAD

The discharge head is usually a cast fitting, provided at the top of the vertical turbine pump column pipes to direct the water flow from a vertical direction to a horizontal direction. It also supports the right angle drive to connect the pump driver to the pump drive shaft (see Figures 9-5 and 9-6).

9.8.3. RIGHT ANGLE GEAR DRIVES

Right angle gear drives are used as power transmission units for the connection of prime movers to pumps. They are used where the horizontal output shaft of the driver must be directed downwards towards a below grade vertical shaft turbine type firewater pump.

They are normally fitted on top of a discharge head at the top of a pump column and its output shaft is connected to the downhole pump. The horizontal input shaft is connected to the driver. A thrust bearing located in the gearbox carries the weight of the

Figure 9-5 Side View of Discharge Head

Figure 9-6 Bottom View of Discharge Head

pump rotating pumps and any unbalanced hydraulic thrust. An anti-reverse ratchet is normally provided to prevent the pump from contra-rotating on shutdown. Hollow shafts are used for gear drives to provide a means for direct connection to lineshaft pumps and to achieve lateral adjustment of the pump impellers.

Currently, gear drives are not required to be approved or listed by NFPA 20, although some manufactures have obtained this listing and it is being increasingly required for firewater pump installations (i.e. Factory Mutual (FM) Approval Standard Class Number 1338). The gear drive should also meet the latest standards of the

American Gear Manufacturers' Association (AGMA) (e.g. AMGA Standard 2003-A86) or German Standards (DIN) for their design and construction to ensure both strength and durability when in service.

Right angle gears used for water pumping applications typically use a service factor of not less than 1.5 (at rated horsepower). The efficiency of a right angle gear drive varies with speed, power and thrust; normally they are about 94 to 98 percent efficient. Due to the rotating elements of a gear drive and need for lubrication, all gear drives should be totally enclosed.

Approval by FM for right angle gear drives normally requires the following information to be submitted by the manufacturer (reference: Factory Mutual Research Corporation (FMRC), Approval Standard, Right Angle Gear Drives, Class Number 1338):

- Strength calculations for the drive shafts and gears (the calculated fatigue load of the gears shall not exceed 50 percent of the actual fatigue strength for the material, based on the maximum load).
- Sample calculations for determining the total pump thrust, shaft size and ball or taper bearing life.
- Detailed drawings of each part used in the drive with material lists and physical property specifications.
- General assembly drawings.
- Maintenance and installation instructions.

When a right angle gear drive is ordered from a manufacturer, the following information is required on the purchase order by the buyer:

- Horsepower requirement.
- Input and output rpm requirements.
- Service factor.
- External thrust.
- Hollow or solid shaft specification.
- Non-reverse capability.
- Instrumentation (gages for lube oil temperature, pressure and level).
- Input and output shaft rotation directions.

(See Table 9-3.)

Offshore gear installations should also be specified as a "marine" package. This generally includes epoxy paint, stainless steel hardware and a copper-nickel (Cu-Ni)

Table 9-3 Shaft Rotation Options for Right Angle Gear Drives

Option	Horizontal Shaft Rotation	Vertical Shaft Rotation
Option 1	Clockwise	Counter clockwise
Option 2	Clockwise	Clockwise
Option 3	Counter clockwise	Counter clockwise
Option 4	Counter clockwise	Clockwise

lube oil heat exchanger. This provides additional resistance to the corrosive effects of salt water.

Gear drives require oil lubrication during their operation. A gear-driven oil displacement pump is usually provided as part of the assembly to provide lubricating and sometimes cooling oil under pressure to all the gears and bearings. Where the operating temperature of the drive may be considered detrimental to the unit and it may be operated more than eight hours per day, a cooling system should be provided for the drive.

The lubricating oil is normally cooled by an oil to water external shell heat exchanger, although air-cooling can be used on some models. Water can be drawn off the firewater pump discharge to supply cooling demands. The discharge of the water-cooling should be visible to ensure that the system has not been plugged or has deteriorated (see Figure 9-7).

Internal cooling coils or external air coolers are options that can be considered when cooling water cannot be provided. Oil temperature, pressure and level gages can be fitted to the unit to further enhance monitoring of the system's performance. An internal oil pickup sump is normally provided with a strainer. Without proper lubrication and cooling, components may overheat, weaken and fail. Where low ambient

Figure 9-7 Photo of External Heat Exchanger for Right Angle Gear Drive (under test at pump manufacture)

temperatures may occur, an automatic coil immersion heater is provided for the gearbox oil, with power supplied from the firewater pump controller. This ensures that the oil does not become too viscous when low temperature conditions occur.

Where a firewater pump, its driver or gearbox can be damaged because of the reverse rotation of the system, a non-reverse feature (i.e. ratchet) is used to prevent this occurrence. The gear drive of the pumping system may be damaged by an accidental shock engagement of the anti-reverse feature. A reverse engagement may be caused by the engine backfiring with the clutch engaged or the firewater pump spinning backwards before the anti-rotation pins engage the ratchet on the gear drive. The firewater pump may sometimes spin backwards; this may be especially true for a long vertical lineshaft pump when the pump is shut off and a high vertical column of water exists and falls back on the pump rotating elements.

A nameplate should be securely fixed to the gear drive indicating the gear ratio for the input and output shafts, its rate speed, the manufacture, model and serial number. Figure 9-8 provides an example of a typical right angle gear drive utilized for firewater pumping service.

Figure 9-8 Typical Firewater Pump Right Angle Gear Drive (cutaway view cross section)
Source: Courtesy of Amarillo Gear Company, reprinted with permission.

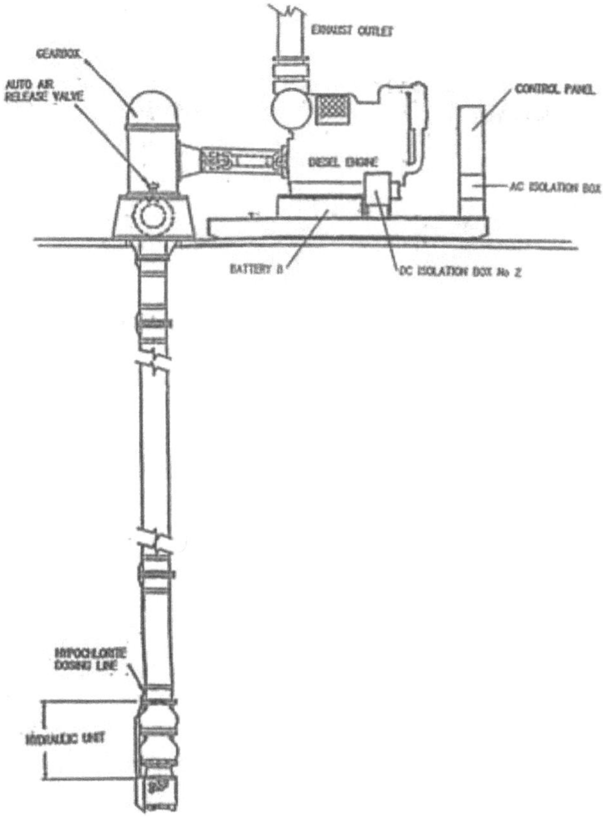

Figure 9-9 Diesel Driven Lineshaft Pump Arrangement

9.9. LINESHAFTS

Lineshafts are the portion of the pump system that connects a well pump to a gearbox. Figure 9-9 shows the general arrangement of a diesel-driven lineshaft pump. They consist of couplings, spiders, bearings and shafts. Screw-type couplings should be avoided, as these may loosen during operation of the pump. The spiders are inserted on the shaft bearings and keep it centered in the pump column. All couplings should be positively locked. The bearings normally require water lubrication, but can tolerate dry running at startup if lined with a suitable material, e.g. PTFE (Teflon) or equivalent. The pump vendor should advise the engine supplier of the required torque to start a lineshaft revolving. Because of the stress and vibration induced in these types of systems, a torsional, vibrational and stress analysis is required. The size of the actual shaft should be in accordance with a recognized standard for deep well pumps, e.g. American National Standards Institute (ANSI)/ American Water Works Association (AWWA) E102-06, *Standard for Submersible Vertical Turbine Pumps*.

FM recommends that "In order to safeguard against the shaft failure, the maximum combined shear stress which occurs in the pump, line or top shaft shall not exceed 30 percent of the elastic limit in tension or be more than 18 percent of the ultimate tensile strength of the shaft material used." (Reference: FMRC, *Approval Standard, Centrifugal Fire Pumps (Vertical-Shaft, Turbine Type), Class Number 1312*, quoted with permission.)

9.10. INDIRECT HYDRAULIC DRIVE

Hydraulic drives are generally chosen for locations where access to the pump is restricted, locations with insufficient room for motors or engines, in locations where the pump can be completely submerged, for locations that are classified and a diesel engine has to sited elsewhere, or in locations where installation of a strictly vertical pump column may be impractical to provide. Hydraulic pumps are therefore most commonly applied to a well, or in particular, an offshore location (see Table 9-4).

Hydraulic pumps are arranged to circulate hydraulic fluid at high pressure. A small hydraulic oil pump drives a remote hydraulic motor which is connected to a submerged lift pump via a short shaft (see Figure 9-10). Double seals prevent leakage both ways. In most cases a two-stage water pumping system is used, with the second stage pump provided at a "topside" location to boost the pressure of a downhole first stage lift pump (see Figures 9-11 and 9-12). Heat exchangers to cool the pumped hydraulic oil are also necessary. In general, hydraulic pumping systems are more expensive than conventional pumping systems, but they offer more options for installation.

Hydraulically driven firewater pumps have found favor with the offshore oil production industry. This is primarily due to the lack of need for an external caisson for the

Table 9-4 Comparison of Offshore Pump Drivers

Feature	Diesel Engine	Submersible Electric Motor	Indirect Hydraulic Drive
Space	For topside engine	Cabling and power source	Topside hydraulic drive and power source
Additional power source needed	No	Yes	Yes
Shaft needed	Long shaft from submerged pump	Very short shaft from submerged motor to pump	Very short shaft from submerged driver to pump
Driver required near pump location	Yes	No	No
Vulnerabilities	Critical alignment and vibration analysis	Vibration and seal leakage to motor	Multiple pumps involved, downhole and booster

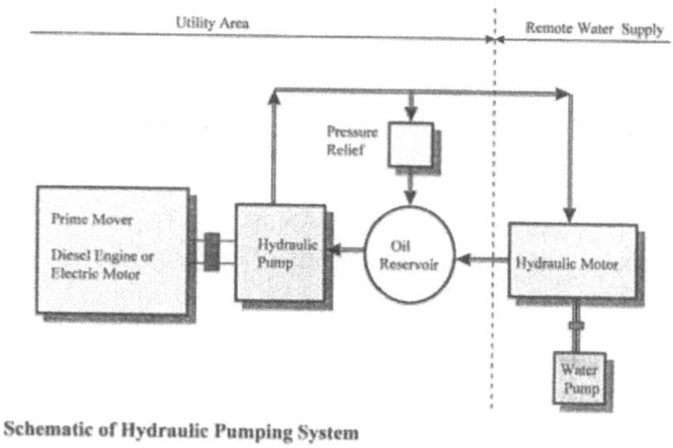

Figure 9-10 Hydraulic Pump System Schematic

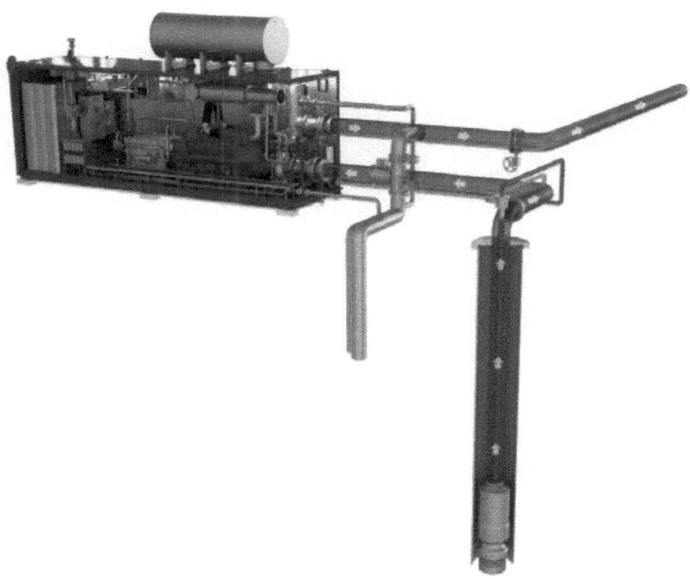

Figure 9-11 Hydraulic Firewater Pump System Arrangements
Source: Courtesy of Frank Mohn Flatoy AS, reprinted with permission.

firewater lift column pipe, resulting in less structural weight and avoidance of the need for an absolutely vertical lineshaft.

Favorable characteristics of hydraulic drive pumps are listed as follows:

- Long shaft is avoided.
- A caisson is not required (an offshore facility consideration).
- Prime mover can be moved any distance away from pump.

- Pump column can accommodate variations to vertical or can even be installed at an angle.
- Right angle gear drive is not required.
- No inertia problems or clutch required.
- Startup torque low compared to lineshaft pump.
- No bearing or shaft misalignment problems.
- Quick installation and removal.
- Condition of seals and bearings can be constantly monitored.
- Weight reduction possible compared to lineshaft pump (offshore benefit).
- All moving parts cooled and lubricated by hydraulic oil at all times.
- Variable speed control from zero to maximum.
- No start-stop problems.

Figure 9-12 Example of Hydraulic Firewater Pump System on Offshore Facility
Source: Courtesy of Frank Mohn Flatoy AS, reprinted with permission.

Offshore experience seems to verify that the starting and stopping of submerged pumps are the biggest stresses to it and the power transmission system. Hydraulic pumping systems reduce this stress. One vendor has also introduced a continuous running circulation pump by a small electric motor in the three kilowatt (four hp) range. The continuous rotation of hydraulic oil rotates the impeller at 50 to 60 rpm. The rotation of the impeller reduces the starting shock and also prevents the build-up of marine growth around the impeller area.

Currently, firewater pumping systems using indirect hydraulic drives have not been tested by US listing and approval agencies (i.e. Underwriters' Laboratories, Inc (UL) or FM), or recognized in the NFPA fire code for firewater pumps, although the drivers and pumps may be independently approved. They have found considerable usage in the North Sea to power firewater pumps because of the advantages they can provide to offshore installations compared to firewater pumping systems using conventional drive transmission systems. In fact, the largest firewater pumps in the world are reported to be installed on the "Ekofisk" offshore oil production complex in the Norwegian sector of the North Sea. Seven 47,332 l/min (12,500 gpm) firewater electro-submersible pumps were provided to the complex in 1991. Each is supplied by its own dedicated diesel generator enclosed by a fire rated module in a topside location.

In general, there are four common arrangements for hydraulically driven firewater pumps. These include the following (schematically illustrated in Figures 9-13A and 9-13B):

1. Vertically submerged lift pump, with a close coupled diesel driver.
2. Vertically submerged lift pump, with a remote coupled diesel driver.
3. Dry mounted lift pump, with a close coupled diesel driver.
4. Dry mounted lift pump, with a remote coupled diesel driver.

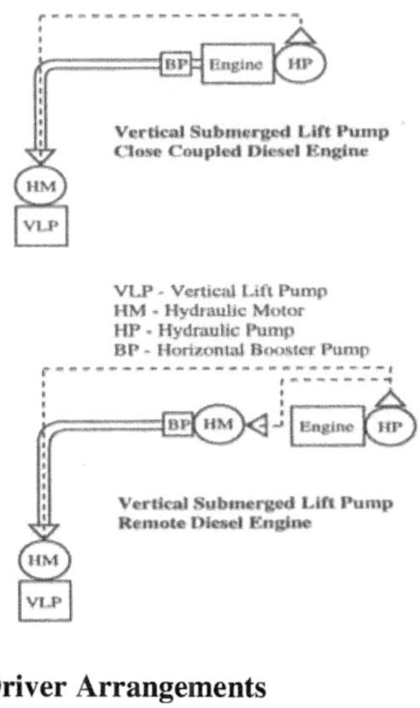

Figure 9-13A Hydraulic Driver Arrangements

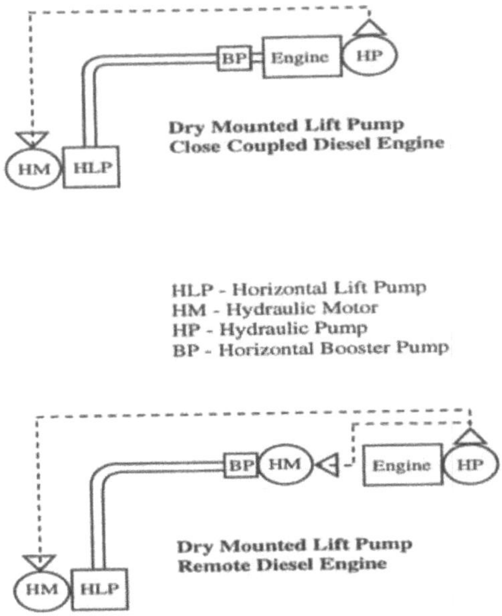

Figure 9-13B Hydraulic Driver Arrangements

9.11. ACOUSTICAL CONCERNS

Where excessive equipment noise cannot be tolerated, it should be eliminated by noise abatement measures. Where noise production is inherent in the equipment, abatement measures may be applied including insulation, noise hoods, operator protective equipment, etc. Noise control measures should not interfere with or obstruct the operation or routine maintenance of a firewater pump or its driver. The vendor of the pump system should supply or state on the pump data sheet the maximum sound level of the equipment or system. API Standard 615, *Sound Control of Mechanical Equipment for Refinery Service*, should be consulted for advice. Chapter 14 discusses further aspects of noise concerns.

9.12. MAINTENANCE ACCESS

One very important aspect, which should be considered in the design and installation of all firewater pumping systems, is access for maintenance and removal of the equipment. Shaft driven systems need sufficient headroom for the easy withdrawal of shafts and columns, and submersible pumps need space to winch or lift the drive units. For offshore facilities, once parts are contained on deck levels, there must be convenient access to the facility crane to move them to a workshop or waiting workboat and vice versa for reinstallation.

Adequate space also needs to be provided around electrical panels for personnel to open doors and avoid electrical shocks.

10 FIREWATER PUMP CONTROLLERS

10.1. INTRODUCTION

The fire main should be brought up to design pressure and flow as soon as hydraulically possible in case of fire. Since firewater pumps pressurize the fire main in the majority of cases, they must be activated as soon as possible during a fire incident to be highly effective.

Firewater pumps are normally arranged to be started manually, or by remote manual or automatic means. They should only be arranged for manual shutdown at the pump itself. Because of their critical importance for startup, automatic startup of the pumps is normally required. Automatic activation is provided through instrumentation connected to an automatic fire pump control box or controller. The controller's primary function is to start up and monitor the condition of the firewater pumping system.

The purpose of the automatic startup of firewater pumps is twofold. First, it ensures the rapid response needed by anticipated firewater demands and provides it immediately. Second, it provides a reliable response that is not dependent on human intervention during an incident (see Figure 10-1).

The purpose of providing remote starting facilities is to enable pumps to be started quickly from a manned central control location, which receives information from other facility safety systems, i.e. fire and gas detection. Remote activation may need to be resorted to if a pump has failed to start automatically or if it is needed urgently, which was previously set in a non-automatic startup mode. While it should always be possible to start a firewater pump from a local control point, this feature should be considered to be a last resort. Considering the time delay involved in personnel arriving at the pump location, local startup during an emergency may not be possible during a serious fire or explosion incident.

Manual startup is provided with an adequately labeled switch provided locally at the firewater pump.

10.2. FIREWATER PUMP CONTROLLERS

Firewater pump controllers perform a vital function in receiving external inputs and sending activation signals to start a pump, monitoring, and in sustaining the operation of the firewater pumping system, in the event of a fire. They are required to provide this support continually throughout the life of the firewater pumping system. Therefore, their reliability is vital to ensure a dependable firewater pumping system is available. Controllers are therefore required to be listed or approved for firewater pumping

Fire Fighting Pumping Systems at Industrial Facilities. DOI: 10.1016/B978-1-4377-4471-2.00010-4

Figure 10-1 Typical Diesel Engine Fire Pump Controller

service through an independent testing laboratory to a recognized standard. They must be independent panels that do not perform any other function except to support the firewater pumping system.

There are two basic types of controller, one for electrically driven pumps and one for diesel engines. Controllers for electrically driven firewater pumps are essentially mechanisms to control the flow or electricity to the motor, while controllers for diesel driven pumps control the startup attempts for the engine and monitor the engine performance.

Controllers for electrically driven firewater pumps can be subdivided into three different types—Full Service Controllers, High Voltage Controllers and Limited Service Controllers. Controllers which are not designed for locked rotor conditions are known as Limited Service Controllers and are used with across-the-line squirrel cage motors of 22 kilowatts (30 hp) or less. High Voltage Controllers are for applications involving 600 volts and above. They are more restrictive for operation and exposure to high voltages, and the Full Service Controllers encompass the remaining applications and are the most common. Full Service Controllers are provided in a variety of types according to the type of motor installed and wiring arrangements employed.

10.2.1. DIESEL ENGINE

Controllers for firewater pumps function as engine condition monitoring and starting devices. They normally contain alarm and signal devices, a pressure recorder and starting and control circuits. See Figure 10-2.

Figure 10-2 Electrical Fire Pump Controller

10.2.2. ELECTRIC MOTOR

Full Service Controllers for electric motors contain several basic common components. These include a voltage surge arrestor, isolating switch, circuit breaker, locked rotor overcurrent protection, motor contactor, pressure recorder (for automatic service) and alarm and signal devices.

10.3. CONTROLLER POWER SUPPLIES

Controllers for firewater pumps should be provided with a reliable power source that will not fail in an emergency. Only one source of power for controller support is normally required, however it is usually prudent to arrange automatic transfer switches for alternative power supplies or dual power feeds, especially when there is doubt as to the reliability or vulnerability of the power supply.

The power supply wiring installation should meet the requirements of the local electrical codes (i.e. National Fire Protection Association (NFPA) 70, National Electrical Code) and that recommended by the fire code for the pump itself (i.e. NFPA 20). It is a common requirement that no disconnection means be provided in the feeders to the firewater pump controllers, to prevent inadvertent shutoff of power to the controller. Exceptions to this requirement are allowed and detailed in the applicable fire codes.

10.4. DUAL POWER SOURCE CONTROLLERS

Firewater pump controllers for electrical motors are normally arranged to receive and control the power source from a single reliable supply. If this supply fails, the firewater pump will be unavailable. Consequentially, controllers are commercially available that can supply power to a motor driven firewater pump from two separate power sources such as normal power and emergency power generation systems. These systems utilize an automatic transfer switch mechanism to transfer to the alternative power source.

10.5. AUTOMATIC TRANSFER SWITCHES

Automatic transfer switches (ATS) are provided to change the power source automatically to a firewater pump from the normally supplied power to an alternative source of power. These switches have sensing devices that monitor each phase of the normal power. When the voltage of any phase falls below the pre-set level, the transfer switch automatically transfers to the alternative source. The switch provides a special circuit that de-energizes the motor control circuit for several seconds prior to the transfer in either direction to prevent high current transients due to an out-of-phase condition between the motor and the source to which it is being connected.

ATS are to be segregated from the firewater pump controller components to prevent the spread of a fault in the circuits. A test switch can be provided to simulate the loss of power so that the transfer switch operation can be checked without interrupting the normal power to the fire pump controller.

10.6. REMOTE ALARM AND SHUTDOWN PANELS

Although each firewater pump is provided with its own local controller, it is highly advantageous to know the status of the firewater pump(s) at a centralized location that is constantly manned. In most cases, remote common signals are provided to the facility utility monitoring station or control room to indicate or alarm automatic firewater pump startup, as a minimum.

10.7. LOW SUCTION PRESSURE CUT-OFF

Low suction pressure cut-off controls are available that will automatically shut down stationary fire pumps that receive water from municipal water supply sources when the water supply pressure reduces to 140 kPa (20 psig) or lower at the suction side of the pump. The panel also features an automatic reset after the water pressure has been restored to above the cut-off limit (after a suitable time delay to prevent pump cycling). It should be remembered that NFPA 20 does not allow the provision of a device installed in the firewater pump suction piping to restrict the starting, stopping or discharge of a firewater pump. Therefore, whenever such devices are considered, their installation should be approved by all concerned parties.

10.8. JOCKEY PUMP CONTROLLERS

In order for jockey pumps to perform as pressure maintenance pumps, a method of instrumentation and control is required, therefore a pump "controller" is required for jockey pumps. In some instances a jockey pump may be arranged to continually operate and in these cases a controller mechanism is not required, only an on and off switch. Because the water supply from jockey pumps is not critical, a listed or approved controller for firewater pump service is not normally required. Controllers for jockey pump service are available that are built to and have been listed to Underwriters' Laboratories, Inc (UL) 508, *Standard for Industrial Controls*. Jockey pump controllers activate from pressure switch inputs. The pressure switch should be located on the system side (i.e. downstream) of the firewater pumps.

10.9. FOAM PUMP CONTROLLERS

Due to the critical need for foam in the suppression of hydrocarbon or chemical fires, foam concentrate pump controllers are required to meet the same requirements as specified for firewater pump controllers.

10.10. CONTROLLER LISTING OR APPROVAL

Most firewater pump installations are required to be in conformance with NFPA 20 and other national or local regulations, and therefore listed or approved controllers are required for fire pump service. This approval ensures reliability and dependability in the performance of its functions. Where controllers are proposed that are not listed or approved, they should be demonstrated to have sufficient reliability to meet the demands of firewater service.

10.11. MULTIPLE FIREWATER PUMP INSTALLATIONS

When more than one pump is installed, they should be coordinated to start in sequence upon further demands for increased water flow requirements. Immediate startup of all pumps may not be necessary and could cause damage to the distribution system. Depending on the number of pumps available, they can be arranged to start up on sequentially decreasing fire main pressure set points.

10.12. FIREWATER PUMP STARTUP

10.12.1. AUTOMATIC ACTIVATION

Automatic activation is the preferred method of firewater pump startup. Automatic activation of a firewater pump may be arranged from a variety of stimuli. These can range from pressure decrease in the fire main to confirmed fire detection—Table 16 provides a listing of some of the most commonly employed automatic startup means. For personnel safety, all firewater pumps that are started automatically or by remote control should have suitable guards and be properly posted. Additionally, all firewater pumps which take suction in areas where underwater diver operations occur, should be provided with suitable guards on the pump intake to preclude the necessity for placing a pump in a non-automatic start mode for reasons of diver safety.

10.12.2. FIRE MAIN PRESSURE SWITCH ACTIVATION

Most firewater pumps are arranged to start automatically when the pressure in the firewater main lowers to a predetermined set point. This is achieved by the provision of a pressure switch fitted on a water line routed to the fire pump controller, or from a switch installed on the fire main itself (i.e. on the discharge side of the firewater pumps). The activation of the pump by a pressure switch is a highly important function, and there have been cases where this has been a single point failure (SPF) for the system. Therefore, some industrial installations provide duplicate or triplicate pressure switches, each with their own independent fire main sensing points. The controller is then provided with logical analysis of the pressure switches, where a two out of three (2oo3) signal confirmation is an indication of fire main pressure drop. This is considered especially important where salt or brackish water sources are used and the pressure switches may be vulnerable to internal corrosion, which would hinder operation. Additionally, these switches should be located where they will not be vulnerable to an incident and are separated from each other to avoid common failure events.

10.12.3. REMOTE ACTIVATION

All firewater pumps should be able to be started from remote activation switches located in a constantly attended location, such as a control room. The remote activation switch should be adequately identified and provided with feedback indicators that show that the pump has started.

10.12.4. LOCAL ACTIVATION

Besides the provision of a controller that automatically starts a firewater pump from various inputs, a local control button is needed to start the pump manually if the need arises. The control button is a simple on and off switch for an electrical motor or a manual push button to activate a starting mechanism for a diesel engine. These switches are commonly incorporated on the controllers themselves.

10.12.5. STARTUP ATTEMPTS

Because some diesel driven firewater pumps may fail to operate on the first starting attempt, NFPA 20 requires controllers to provide six attempts at starting of 15 seconds each, separated by dwell periods of 10 seconds. These six attempts are continued until they are canceled by operator intervention or the controller receives an engine running signal.

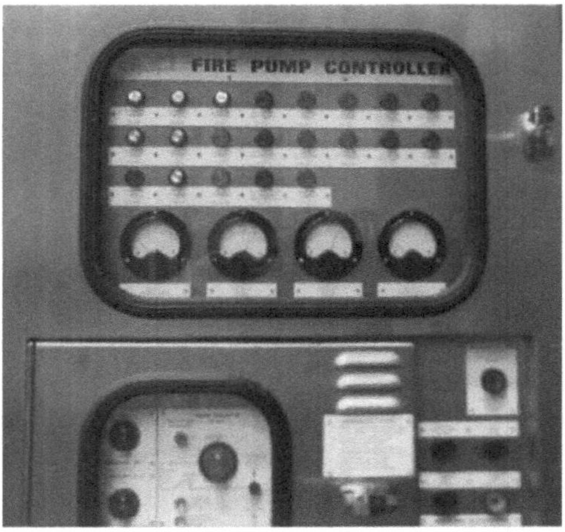

Figure 10-3 Diesel Engine Controller Panel with Lamp Display

10.13. COLOR CODING OF PANEL INDICATORS

The indicators on a controller panel should be color coded to readily indicate the condition of the feature being monitored. Usually, colored lenses or lamps are fitted for this purpose (see Figure 10-3). NFPA 79, *Electrical Standard for Industrial Machinery*, provides guidance in the provision indicator lamps and colors commonly provided. Although this is commonly used, it is not a rigid feature adhered to by all vendors. Most controllers are now Programmable Logic Controller (PLC) based with LED readouts which have supplemented the use of colored lamps or indicators for showing the condition of the system.

10.14. ELECTRONIC READOUT DISPLAYS

The latest trend for fire pump controllers is to have messages of the fire pump operation display in alpha-numeric characters in a display. Depending on the quality of the controller selected this can be as minimal as a coded reference number or actual messages to indicate the exact condition of the pump, i.e. operation conditions.

10.15. PIPING AND INSTRUMENTATION DIAGRAMS

Engineering drawings prepared for the process industries for the provision of pump arrangements are provided on Piping and Instrumentation Drawings (P & IDs). These drawings show the schematic arrangement of the process piping, valves, vessels, and instrumentation points for control and measurement. Actual arrangements of piping and structural supports are shown on piping isometric drawings and structural detail drawings. Because the actual arrangement drawings indicate the features and

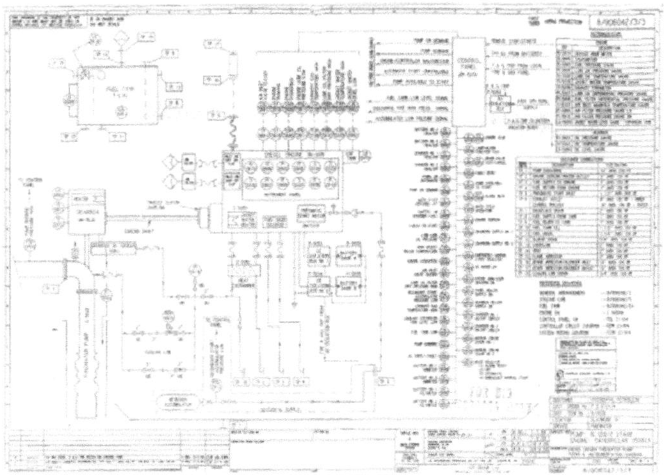

Figure 10.4 Typical P & ID for Offshore Firewater Pump Installation
Source: Courtesy of Occidental Petroleum Corporation, reprinted with permission.

interconnections for a pump, they are the prime drawing for the specific installation of a pump or other process equipment. The same holds true for firewater pumps. Because of their interconnections with other shutdowns, alarms and other emergency and support systems, the arrangements that are required may become quite complex.

Usually P & ID and General Arrangement Drawings are prepared for a proposed pump installation. Once the P & ID is approved, detailed piping isometrics, electrical interconnection diagrams and structural details can be prepared. A Fire Protection Engineer (FPE) should review P & IDs for firewater pump installations to ensure compliance with pertinent fire code regulations and company standards. Additionally, some localities require that arrangements for firewater pumps be submitted to governmental agencies for approval. Figure 10-4 provides an example of a P & ID that was prepared for an offshore platform firewater pump installation on an oil and gas production platform in the North Sea.

10.16. CONTROLLER INDICATORS

Controller indicators can range from the basic minimum to the very extensive. The types of indicators desired should be defined as part of the purchase order for the pump unit. As firewater pump systems become more complex, more indicators are provided to control and monitor the condition and performance of the system. A common alarm and trouble signal is normally sent to a main installation control point from the firewater pump controller. The alarm signal indicates pump startup and failures on the system.

Because fewer components are involved with electrically driven firewater pumps than diesel driven pumps, these have considerably less items to ennunciate.

Electrical Motor Controllers: Required to have as a minimum the following alarm indicators with activation of an audible alarm:

Power Available
Phase Reversal.

Diesel Engine Controllers: Required to have as a minimum the following alarm indicators with activation of an audible alarm:

Controller in "Automatic" Mode
Low Engine Oil Pressure
High Water Temperature
Failure to Start Automatically (overcrank indication)
Overspeed Shutdown
Battery Failure (one for each battery set)
Battery Charge Failure (one for each battery set charger, also indicates loss of alternating current power to charger)
Low Air or Hydraulic Pressure (when provided).

Additionally, an ammeter is required for a diesel engine controller to indicate the battery charger's rate of charge.

The minimum required and some optional indicators for controllers are indicated in Table 10-1, along with the preferred color. NFPA 79, *Electrical Standard for Industrial Machinery*, provides guidance in the color of the indicators that should be used.

10.17. FIRST-UP FAULT FEATURE

Where fault features are provided on a control panel, to indicate the failure of the system, all the failure indicators may have activated by the time someone arrives at the panel. In this case, it is helpful to have a first-up fault feature that indicates which fault occurred first, to aid in troubleshooting or correcting the difficulty. Microprocessor control systems which are arranged to monitor utility systems can log when trouble or alarm signals are received. Where separate signals are fed into these systems from fire pump controllers, the nature of pump failures may be readily identified.

Microprocessor-based fire pump controllers are slowly appearing on the market, and may contain some of the features that are similar to fire alarm panels which are microprocessor-based, and record detection and alarm actions along with a date and time stamp.

10.18. CAUSE AND EFFECTS CHARTS

Cause and effects charts are prepared for a firewater system to determine what actions are to be taken by the system in response to various inputs. It is essentially a logic chart

Table 10-1 Diesel Engine Controller Indicators

Indicator Type	Minimum Required	Optional Feature	Preferred Color
System in Automatic	X		Green
Low Lube Oil—Trouble	X		Red
High Engine Coolant Temperature—Trouble	X		Red
Failure to Start	X		Red
Overspeed	X		Red
Battery A Fault	X		Red
Battery B Fault	X		Red
Charger A Fault	X		Red
Charger B Fault	X		Red
Starting Air Supply Failure	X		Red
Pump Available		X	Green
Pump on Demand		X	Yellow
Pump Running		X	Yellow
Engine Running		X	Yellow
Automatic Start Switched Off		X	Red
Battery A Healthy		X	Green
Battery B Healthy		X	Green
Charger A Healthy		X	Green
Charger B Healthy		X	Green
Lube Oil Filter Differential High Pressure		X	Red
Exhaust Gas High Temperature		X	Red
Coolant Header Tank Low Level		X	Red
AC Mains On		X	Orange
AC Supply Fault		X	Red
Low Fuel Level		X	Red
Low Water Supply		X	Red
Relief Valve Open		X	Red
Engine Immersion Heater On		X	Orange
Panel Heater On		X	Orange
Gearbox Oil Heater On		X	Orange
Air Flaps Closed		X	Red
Lamp Test		X	Black Button
Alarm Silence		X	Yellow
Manual Start Selected		x	Red

to indicate inputs and outputs. Table 10-2 provides an example of a simple Cause and Effects chart for the operation of a firewater pump. All the inputs, or causes, are listed on one side of the chart and the effects, or resultant actions of these inputs, are listed cn an adjacent side. Crosses ("X") or check marks ("⌦") are placed on the chart blocks

Table 10-2 Cause and Effects Chart—Firewater Pump Startup

Effects Causes	Startup of Jockey Pump	Startup of Main Firewater Pump(s)	Startup of Backup Firewater Pump(s)
Minor Drop in Firewater System Pressure	X		
Major Drop in Firewater System Pressure		X	
Continual Drop in Firewater System Pressure			X
Main Firewater Pump Fails to Start			X
Main Firewater Pump Fails during Operation			X
Firewater Pump Controller Sequence Starter Activation		X	X
Request for Firewater Pump Startup from Control Room		X	
Request for Firewater Pump Startup from Remote Location		X	
Activation of Plant ESD System		X	
Confirmed Fire, Heat or Smoke Detection		X	
Confirmed Combustible Gas Detection		X	
Activation of Fixed Firewater Suppression System(s)		X	

where a cause has been programmed to effect a certain action. By this fashion, an understanding of how a system will perform under various conditions can be made. Where special set points activate the system, these can also be noted on the chart.

The programming and inputs to all firewater pump controllers should be based on a Cause and Effect Chart that has been agreed upon by the operational staff with input from a Fire Protection Engineer. This allows the determination of the best arrangement that is suitable for the facility, and which will also meet life safety concerns.

10.19. FIREWATER PUMP SHUTDOWN

Once a firewater pump has been activated, it should not be automatically shutdown. Supplemental water pressure from mobile or other sources during an emergency may provide adequate pressure to the fire main, which may be misinterpreted by automatic systems that the situation is normal. Although NFPA 20 does allow automatic shutdown if all the startup signals have returned to normal, however, as previously pointed out, these signals may be misleading if water pressure is the sole indicator for pump startup. Additionally, onsite physical verification of the fire incident control should be accomplished before firewater pumps are inadvertently shut off. Engine overspeed may be a condition that allows a power to be removed from engine running devices which may cause the engine to be eventually shutdown.

The manual shutdown capability should only be provided at the dedicated controller for the firewater pump that is located within the immediate vicinity of the unit (commonly on the local pump controller). Some installations have provisions in a control room for the manual or automatic modes for firewater pumps. Serious consideration should be given to avoid the provision of automatic over-ride, because if the firewater pump is placed in manual and the control circuit is damaged in an incident, it subsequently will not be able to be started up remotely (i.e. switched back to automatic) and immediately support the incident.

One of the faults found in the Piper Alpha Disaster in 1988 was that the firewater pumps had been switched to a manual mode (i.e. needed to be started locally), therefore the possibility of automatic startup during the fire incident was negated. The subsequent inquiry ("Cullen" Report, Recommendations 51 and 70) recommended that firewater pumps in particular be examined to determine their ability to withstand severe accident conditions, and the operator should set acceptance standards for its availability. NFPA 20 recommends that only remote startup be provided to firewater pumps, not remote stopping.

10.20. SPECIALIZED INSTALLATIONS

On jack-ups and some semi-submersible rigs, the firewater pump suction is typically fed from a "raw seawater system" (which is primarily used for buoyancy control and drilling operations). The fire main is normally kept at the supplied pressure of the raw seawater system. When required, the rig firewater pumps boost the pressure of the firewater system from the raw seawater supply system.

Because there is a high demand with large fluctuations for raw seawater usage in these facilities (i.e. for drilling and buoyancy), the firewater pumps are not controlled by pressure drops in firewater main. Special actuating arrangements are therefore provided for these installations, such as immediate remote activation from a continuously manned control room, startup upon confirmed fire detection, activation of a water system, etc.

10.21. CONTROLLER LOCATION AND ACCESS REQUIREMENTS

The controller should be provided in the general vicinity of the firewater pump which it operates. Common practice is to mount the controller on the same skid that the firewater pump is located on. Access to the controller should be available from two different locations. Where the controller is provided at a multi-level facility, it should be located near the stairwell for rapid and convenient access by emergency personnel.

Access for working on the controller itself should meet local electrical code requirements, commonly that of NFPA 70, *The National Electrical Code (NEC)*, article 110.

11 RELIABILITY

11.1. INTRODUCTION

Failure of the firewater pumping system has been cited as a major contributor to 12 of the 100 largest petroleum industry incidents and ensuring fires, therefore reliability of the firewater system is a very critical feature. Failures can be caused by mechanical or electrical failures, or because of an impact from the initiation of the fire incident (i.e. blast overpressure). This chapter is primarily concerned with the mechanical and electrical reliability of the firewater pumping system.

First, one has to ascertain the definition of reliability as applied to firewater systems. Reliability is the probability that an item is able to perform a required function under stated conditions for a stated period of time or for a stated demand. There are several components that make up a firewater system (i.e. supply source, pumps, distribution piping, etc.) and each has its own reliability rate. In our case, we are particularly interested in the reliability of the firewater pump itself. In some cases, it is worthwhile to conduct a firewater reliability analysis (FRA) of the entire firewater system to determine the risk of failure.

In a broad sense it can be stated that the reliability of a firewater pump is the probability that it will perform successfully in the event it is called upon to perform during an emergency. The key wording is "perform successfully". In this it is meant that it will start up as required and deliver the required firewater flows and pressures for the specified duration of the emergency. The reliability of the firewater pump can therefore be established by examining the failure rates of its various components and its redundant features.

Fire pumps should be expected to operate on intermittent duty (i.e. one hour per week for 10 or 20 years) or for continuous duty (i.e. 1,000 to 10,000 hours), whichever is considered the more severe or realistic application. National Fire Protection Association (NFPA) 20 requires that vertical shaft turbine pumps have a thrust bearing that is constructed with an average life rating for five years of continuous operation (i.e. 43,800 hours). Although there is no specific requirement to run a fire pump continuously for extended periods (except when an industrial facility is undergoing critical operations or startup and requires fire pumps online), it does represent the maximum operational condition that it could be expected to experience during its entire lifetime. By examining these periods of operation, one can envisage the quality of pump design and reliability that is required for the unit.

One of the prime features that influences the reliability of mechanical and electrical devices is adequate and timely maintenance and service. Firewater pumps must start instantly after a period of standing idle.

The basic need for a reliable firewater system is primarily economic. It must be determined whether the firewater system has adequate reliability to reduce economic

Fire Fighting Pumping Systems at Industrial Facilities. DOI: 10.1016/B978-1-4377-4471-2.00011-6

losses due to potential fires (although social impacts must also be considered to some extent).

11.2. FAILURE CATEGORIES

Failures associated with equipment can be categorized into two broad areas—generic failures or specific operating circumstances. Generic failures can be caused by failure of each mechanical component, corrosion, vibration, or external impact. Specific failures are related to operation circumstances and are primarily related to human error.

11.3. INSURANCE INDUSTRY EXPERIENCE

A published study conducted by Marsh & McLennan Protection Consultants (M&MPC) was reported in 1990, on the ability of a firewater pump to perform adequately, with half of those tested located in the hydrocarbon and chemical industries. They found that, out of the 400 pumps tested, 38 percent of the fire pumps provided for hydrocarbon or chemical industries failed, compared to 17 percent for other industries. The failure was defined as water delivery "below 90 percent of the pump's rated capacity at the time of the test".

11.4. FAULT TREE ANALYSIS

A fault tree is a risk analysis method describing how a plant hazard or other undesirable event may occur in terms of combinations of individual non-hazardous component or operator failures. The fault tree evaluates the probabilities of these failures occurring in a tiered manner of combinations which may be obscure at face value (see Figures 11-1 and 11-2). It can be used to identify possible system failures, predict reliability, availability or failure frequency, identify system improvements, predict the effects of changes on the system, and also can be used to understand the operation of the system. As a firewater pumping system has several components, it is an ideal candidate for a fault tree analysis (FTA).

11.5. SINGLE POINT FAILURES

In most equipment where reliability is desired, duplication of some parts will be necessary. This need is dependent upon the extent to which the various parts may reasonably be expected to be out of service as a result of maintenance and repair work, an emergency or other unknown condition. If a single part or point in the system affects the entire operation, it is considered to be a single point failure (SPF) for the process. It will therefore warrant special investigation on whether its failure rate can be considered

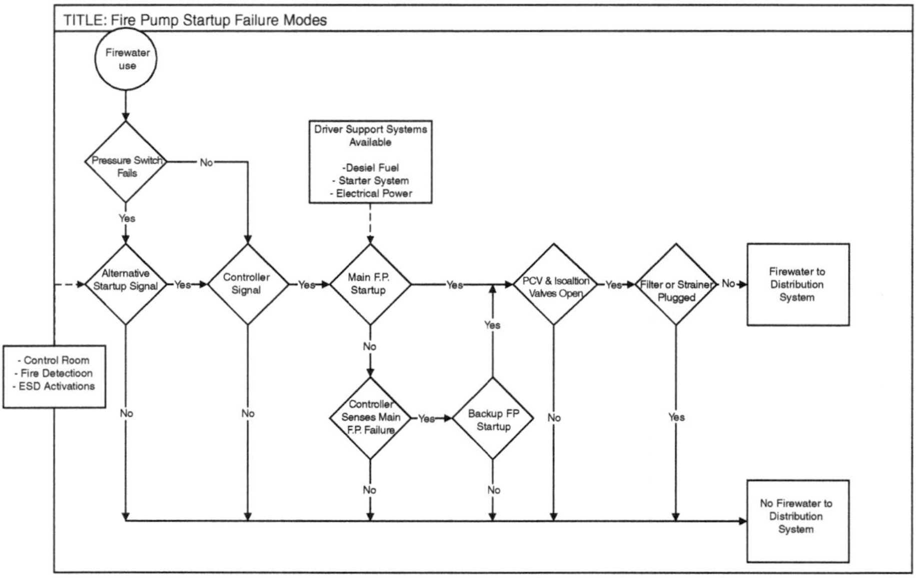

Figure 11-1 Flowchart of Firewater Pump Failures

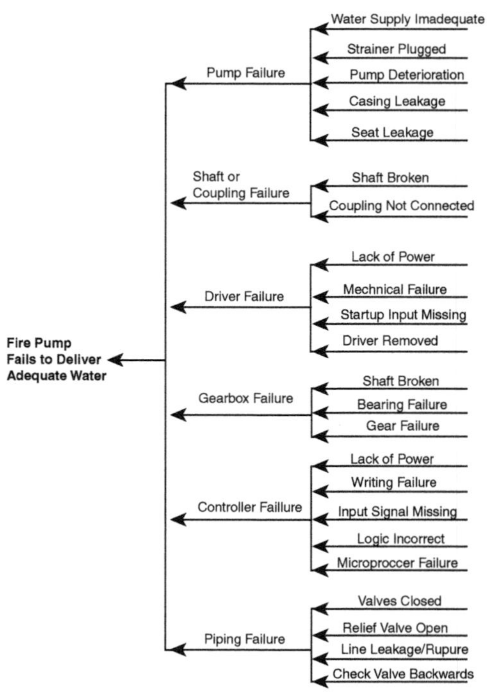

Figure 11-2 Firewater Pumping System Failure Modes

as frequent and therefore whether duplication is necessary. Firewater pumping systems should be specifically examined to determine if SPF points exist in the system.

11.6. NUMBER OF FIREWATER PUMPS

The minimum number of firewater pumps provided for an industrial facility is usually two; a main pump and a backup pump in case the main pump fails to operate (i.e. two 100% pumps). There may be times when one of the pumps, either the main or backup pump, is removed for maintenance or refurbishment. This leads to the situation where only one fire pump is available for support, which could fail upon startup. As industrial facilities are continuously operated and cannot be allowed to operate without firewater support, they must shut down immediately if the remaining firewater pump fails because of the high risk involved. As a shutdown of production is not desired, a third backup pump is normally required, especially in cases where a production shutdown is an undue economic burden (i.e. loss of cash flow) and reduces the probability of a firewater pump failing to startup. Some national regulations now require the provision of three 100 percent firewater pumps for offshore installations, not primarily for production concerns, but because of the concern for life safety during a fire situation (Safety Notice 10/89, Department of Energy, UK) and the difficulties in assured evacuation.

For onshore facilities, a continuously manned fire truck pumper could be temporarily connected to the firewater system as backup for the duration of the fixed pump removal if it is of short duration. Due to the inconvenience of a manned standby crew for fire pump backup support, a third 100 percent firewater pump is regularly provided. Offshore installations do not commonly have the advantage of mobile firewater support, and therefore critical installations require a minimum of three 100 percent pumps (although the author has engineered arrangements where backup firewater support to a fixed offshore structure was provided from a semi-submersible vessel via a catenary hose arrangement as a temporary measure prior to the installation of a third firewater pump). Alternatively, four 50 percent pumps may be provided so that if one unit is removed and one fails, two 50 percent units are still available to provide the required 100 percent support requirements.

11.7. PUMP OPERATIONAL FAILURES

Pumps require regular running to ensure operability. Pumps commonly fail due to corrosion, deterioration and plugging. Periodic performance testing ensures operability and can predict deterioration of pumps due to wear and corrosion activity.

11.8. ELECTRICAL MOTOR FAILURES

Taken on their own, motors are more reliable than engines due to the smaller number of components involved. Motors may fail due to bearing wear or lack of lubrication, wiring faults or improper stresses on the rotor shaft.

Electro-submersible pumps may suffer from vibration, motor seal failure, cable failures and erosion or corrosion of the pump.

Overhead or exposed powerlines to motors are more susceptible to damage and interruption than buried lines, and introduce a degree of unreliability due to their location and construction. Weather conditions and incident exposures can have a damaging effect on the system. The possibly of power failures affecting a large area should also be addressed.

11.9. DIESEL ENGINE FAILURES

The diesel engine is the most dependable of the internal combustion engines. Failure of diesel engines is primarily due to ancillaries; the fuel, lubrication oil, cooling or starting systems. Common engine failures result from contaminated fuel, clogged fuel lines or oil filters, lack of adequate starting power, wiring failure and metallurgical fatigue. Oil changes and overhauls are required periodically, based on the use of the unit. Cleaning of fuel filters and removal of water from traps is vital to avoid breakdowns.

Experience has shown that a diesel engine-driven fire pump is the most reliable in severe loss incidents. Electric, steam, or spark-ignited engines are more likely to be put out of action because of the initial fire or explosion during an incident.

11.10. GEARBOX FAILURES

The lubrication and cooling systems of gearboxes are the primary cause of integrity concerns. Without proper lubrication and cooling systems, components may overheat, weaken and fail. Bearings are generally designed for a useful life of 5,000 hours of continuous operation under maximum load.

11.11. CONTROLLER FAULTS

A study conducted by the UK Health and Safety Executive (HSE) on control system-related accidents, indicates that almost half were caused by incorrect specifications (i.e. 44.1 percent). Table 11-1 summarizes their findings.

Table 11-1 Failure Causes of Control System-Related Accidents

Failure Cause	Percentage
Specification	44.1%
Changes after Commissioning	20.6%
Operation and Maintenance	14.7%
Design and Implementation	14.7%
Installation and Commissioning	5.9%

Controller failures can occur because of loss of power, failure of control boards, relays, instrumentation and improper programming of logic functions.

11.12. PLANT PERILS AND PUMPING SYSTEM EXPOSURE

The primary concern in the reliability of a firewater system during an incident is that it is not itself affected by the incident. In particular, an Unconfined Vapor Cloud Explosion (UVCE) has been shown to impact on the firewater supply where its arrangement, routing or location is vulnerable to an explosion.

12 CLASSIFIED AREA PUMP INSTALLATIONS

12.1. INTRODUCTION

Installation of fire pumps in electrically classified locations is sometimes required when limited spacing or a non-classified area is not available at a chemical or hydrocarbon installation.

This is particularly true for some offshore installations where no other alternative might be available in retrofit installations, or where economic trade-offs for space-weight occurs during the design of the structural platform because of the overall commercial viability of the project. Originally, the demand for classified firewater pump installations was requested from oil companies which were finding the construction of a purge-pressurized enclosure costly and inconvenient, as well as imposing a weight implication. They were also dissatisfied with the complicated ducting necessary for obtaining "safe" air and venting the exhaust to a safe area. These arrangements necessarily interfered with the layout of other equipment on the installation.

There may also be a risk philosophy adopted for a facility that, although the normal limits of area classification by electrical codes are recognized, the operators wish to extend these limits to considerable distances because of the nature of the facility, the risks involved and the desire to ensure fire pump operability in unusual circumstances. Although these pumps are located in normally non-classified areas, they are thought of as being able to operate in classified conditions (i.e. vapor releases) when suitably modified. In this fashion, they may increase the safety margins at the facility. Only a portion of the facility firewater pumping capacity is allowed to be installed in these circumstances, and this is normally only when 300 percent pumping capacity is required (compared to the standard 200 percent in non-classified environments, 100 percent or less in classified environments).

Because of the hazardous environment a pump faces in these locations, they are more liable to fail unless special precautions are taken (failure can range from explosion effects to the system from the vapor in the areas igniting, overspeed failure from ingestion of vapors, or possibly internal explosion of the driver itself from the ingestion of vapors). In fact, some insurers recommend against the installation of a firewater pump in a classified area. A fire pump also provides numerous ignition sources for combustible or flammable vapors that must be protected against before a fire pump is allowed to operate in a classified area. Both electrical motors and diesel engine drivers pose ignition hazards that should be recognized and dealt with. Management must be advised of and acknowledge these risks before the provision of a firewater pump in these areas is allowed.

Fire Fighting Pumping Systems at Industrial Facilities. DOI: 10.1016/B978-1-4377-4471-2.00012-8

The most common approach is to construct a firewater pump assembly in a self-contained enclosure (with suitable fire and explosion resistant barriers) where the interior is a safe area, but is placed within the classified area. Air is drawn into the enclosure from a non-classified area to maintain the non-classified status with appropriate vapor transmission controls for personnel entrances (i.e. airlocks). This allows for the provision of an ordinary firewater pump in a classified area without any modification and less probability of failure. This option carries with it the extra cost, weight and space of the enclosure and additional services to support it. The alternative to this approach is to construct the firewater pumping system (specifically its drivers and controllers) to meet the classified area requirements. In some cases both the enclosure and classified pump construction are provided to meet reliability requirements and provide a higher level of protection against accidental ignition from the pumping system.

12.2. DIESEL ENGINE IGNITION HAZARDS

Diesel engines contain some inherent ignition hazards, which can be classified into two general areas—primary and secondary hazards, as identified below. Further information is provided concerning the elimination or reduction of these hazards later in this chapter.

12.2.1. PRIMARY IGNITION HAZARDS

These hazards can be expected during the normal operation of a diesel engine, which has not been modified to operate in a classified area. Suitable controls and preventive measures can be provided to reduce the primary ignition hazards for diesel engines.

Primary ignition hazards are generally considered to be as follows:

- Engine surface and exhaust gas temperatures.
- Discharge of sparks from the engine exhaust system.
- Overspeeding of the engine.
- Discharge of sparks from engine electrical equipment.
- Flashback through engine air intake system.
- Static electricity discharge.
- Flame transmission through engine decompression ports.

12.2.2. SECONDARY IGNITION HAZARDS

These secondary ignition hazards may be expected as a result of engine or equipment malfunctions, and so have a lower probability of occurring.

Secondary ignition hazards are generally considered to be as follows:

- Discharge of sparks from engine mechanical causes.
- Overheating due to cooling water or lubrication oil failure.
- Excessive engine vibration.
- Explosion in engine intake or exhaust system.
- Explosion in engine crankcase.

12.3. HOT SURFACES

Although there is a general consensus within the industry that hot surfaces are of concern where combustible vapors are released, because they may be an ignition source, there is not a clear position for the limits of hot surfaces that should be used. The National Fire Protection Association (NFPA) has recognized the hazard but has not provided further guidance or regulation regarding the matter. The American Petroleum Institute (API) RP 14C, *Recommended Practice for Analysis, Design, Installation and Testing of Basic Surface Safety Systems for Offshore Production Platforms*, and the UK Institute of Petroleum (IP) *Area Classification Code for Petroleum Installations*, recommends that a limit of 250 °C (482 °F) be used. It is felt that this limit is primarily required to restrict engine temperatures below the ignition temperatures of the more commony encountered flammable gases or vapors. However, API PSD 2216, *Ignition Risk of Hot Surfaces in Open Air* recommends that "ignition by a hot surface should not be assumed unless the surface temperature is at least 200 °C (392 °F) above the normally accepted minimum ignition temperature". There is some evidence to show that smaller surfaces are of lesser concern than larger ones that are at the same elevated temperature. The US Mineral Management Service (MMS) has recorded several incidents in the Gulf of Mexico where vapors were ignited as a result of contact with the exhaust systems of diesel engines.

The surfaces that may be of concern for firewater pumps are the exhaust manifold, mufflers, exhaust piping and exhaust gases from diesel engines. Whatever limit is chosen for the surface temperature, it should be applied under maximum torque conditions of the operation of the engine. The engine itself should be sited or arranged to have free movement of air over its exposed surfaces.

12.4. HOT EXHAUST GASES

Diesel engine exhaust gases vary with speed and load. High loads and high speeds result in the highest temperatures. Generally, temperatures of 500 to 700 °C (932 to 1,293 °F) are produced in the exhaust gases from diesel-cycle engines at 100 percent load, to 200 to 300 °C (392 to 572 °F) with no load. Exhaust gases normally discharge at a temperature of around 420 °C (788 °F).

Many incidents have been recorded where the exhaust of a diesel engine has ignited vapors from a leak. Several methods are used to prevent the hazards of exhaust gases acting as an ignition source. The gases may be directed to a location that is considered safe, or water sprays may be provided in the exhaust piping to lower their temperatures. Exhaust gases should never be directed towards combustible constructions.

12.5. EXHAUST SYSTEM (MUFFLER)

Diesel engine exhaust systems often reach a surface temperature of 350 to 600 °C (662 °F to 1,112 °F), which may be capable of igniting flammable substances. The exhaust system of diesel engines which are required to operate in an electrically classified area must therefore be modified so they will not act as an ignition source.

The exhaust system can be modified so it is a dry insulated system (double skinned—insulated) or a seawater-cooled/jacketed system to keep the exhaust system surface temperature and that of exhaust gases from the diesel engine itself below that required for ignition of released combustible gases in the area.

12.6. EXHAUST SYSTEM SPARK OR FLAME DISCHARGE

On occasion, sparks or flames can be emitted from the exhaust systems of internal combustion engines. To prevent this occurrence for engines operating in a classified area, "spark arrestors" (cyclone or baffle type) are fitted to the discharge of the exhaust system piping. Spark arrestors should be certified by the manufacturer for the correct size and proper arrangement. A test certificate should be provided indicating the effectiveness of the spark arrestor.

12.7. ENGINE OVERSPEEDING

Because of the diesel engine ignition compression cycle, if a flammable atmosphere is present, the engine may continue to run even when normal fuel supply is shut off. It is also commonly thought that if the supply of both normal fuel and ingested fuel from accidental vapor releases is provided to the air intake of a diesel engine, it may possibly cause the engine to overspeed.

In practical application, it is extremely unlikely that a diesel engine will overspeed due to the induction of flammable gases. This is because the power demanded by the pump increases by the cube of the speed increase. In addition, the inducted flammable vapor cannot add increased usable calorific valve to the fuel unless it is introduced to the cylinders under pressure and with increased amounts of oxygen to enable it to burn.

The most likely cause of an engine overspeed would be the failure of the engine governor. This overspeed should be detected by the engine speed sensing device which should shut off the fuel supplies and activate an inlet air shutoff valve as described below. These precautions are recommended by NFPA 37, *Standard for the Installation and Use of Stationary Combustible Engines and Gas Turbines.*

If overspeeding occurs and it is above a normal maximum governed speed, a shutdown valve can be fitted in the air intake to automatically isolate incoming air to the engine air inlet port and stop the engine. The air inlet closure must provide sufficient shutoff of air intake to stop the engine, whether it is operating on normal fuel supplies or possibly due to some ingested vapors. The device is fitted as close to the air intake of the engine as possible and is normally provided between the flame trap and air intake filter. Additionally a manual capability to provide rapid stopping of the engine is fitted to the air intake shutdown valve.

The engine itself should be undamaged because of the overspeed shutdown. Resetting of the air inlet damper would be required, prior to the restart of the engine, following removal of the overspeed cause.

12.8. FLASHBACK IN AIR INTAKE

Air supplies to the engine should be drawn from a safe area. This is to prevent a fuel mixture which is too rich from entering the air intake and causing oxygen starvation, and to prevent ignition flashback to the classified area through the air intake system. Gas detection should be installed at the air inlet to alarm at a manned location. The detection system should alarm at a low point and cause an engine shutdown at a high level (commonly 20 percent lower explosive limit (LEL) for low level and 50 percent LEL for high level).

On occasion, a flashback can occur through the air intake system from the combustion chambers. A flame trap of the crimped, coil metal cartridge type is commonly provided to prevent the passage of flashback to the outside. The flame trap should be easily reachable for inspection and maintenance purposes.

12.9. MATERIAL SELECTION

All external parts, e.g. fan blades, drive belts, etc. which may come into contact with other parts should be manufactured from non-sparking materials. Additionally, all drive belts should be made from antistatic materials and be fire resistant.

12.10. RATED INSTRUMENTATION AND ELECTRICAL HARDWARE

All instrumentation and electrical devices selected for use and supplied by vendors must be certified for use in the intended classified area. Equipment that is not rated for such use should be isolated and shut down upon the confirmed detection of combustible gases or vapors in the area. However, those items required for pump operation should be suitably rated. All electrical equipment should also be grounded or bonded to avoid the generation of sparks.

Since batteries can cause sparking, they should not be used for classified area operations. Instead, air or hydraulic starting systems are provided. If batteries are provided as the primary starting mechanism, they should be automatically disconnected following the detection of flammable gas in the area.

12.11. DECOMPRESSION PORTS

Flame transmission to the atmosphere can occur if decompression ports are provided on the engine. Therefore, decompression mechanisms should not be provided.

12.12. ELECTRIC MOTORS

Electric motors may spark during their operation, therefore they must be certified as being suitable to operate in a classified area. The actual rating should be consistent with the worst ignition commodity the facility handles (i.e. lowest ignition temperature, vapors or fibers, etc.). They should also be adequately bonded or grounded.

12.13. CONTROLLERS

Controllers for firewater pumps contain electrical devices that may be an ignition hazard. They are commonly provided with a purging mechanism to avoid the entry of flammable vapors to the panel, or the panel itself may be constructed to an explosion-proof design.

13 FIREWATER PUMP ACCEPTANCE AND FLOW TESTING

13.1. INTRODUCTION

A method of testing firewater pumps should be provided to verify that adequate flow performance and acceptable mechanical operation of a firewater pump will be available during an emergency. Additionally, most facility fire protection audits, insurance surveys and evaluations and local maintenance requirements would require that firewater pumps be routinely tested for performance verification due to their critical support.

Performance variation greater than 10 percent from the initial field acceptance curve or 15 percent from the manufacturer's curve indicates that corrective action is needed for the pump. It is normal for a field generated pump performance curve to be as low as 90 percent of the factory pump curve due to differences in test methodologies, the stricter tolerances of the factory, use of laboratory tested equipment, etc. Even in highly accurate test procedures for firewater pumps (i.e. for listing or approvals), several +/- percentage variances for various instrumentation and hardware components in the test arrangements are allowed.

In fact, preventive predictive failure maintenance can be performed before firewater pumps reach reduced flow performance levels. The flow performance can be graphed from year to year from the factory curve and a predictive trend can be extrapolated and plotted which will generally indicate the useful life of the unit.

Pressure gages should be provided on the suction and discharge and a method to measure the amount of flowing water produced by the pump. The sizing of flow test piping should account for the maximum flow rate of the unit, not just its rated capacity. A flow test re-circulation loop with pressure and flow indicators or recorders is recommended for conducting regular pump performance tests easily. Where a test loop is not available, pitot tube measurements can be made from straight steam water nozzles from the total water flow from the pump. If nozzles are used, ground surface erosion damage from high water pressure should be avoided by the provision of fittings to allow adjustment of the nozzle elevation and orientation to disperse the water spray.

13.2. SAFETY PRECAUTIONS

Firewater pumps can produce high water pressures and capacities that can inflict fatal injuries and extensive property damage if not properly controlled and managed. All personnel involved in firewater pump tests should be briefed on the associated hazards.

Fire Fighting Pumping Systems at Industrial Facilities. DOI: 10.1016/B978-1-4377-4471-2.00013-X

This is especially important when temporary or portable equipment is used (e.g. hoses and test nozzles) or the pump is operated outside its normal operating range. Temporary or portable equipment may not be properly secured, or the installation itself may not be able to sustain the maximum operating pressure from the firewater pump due to unforeseen circumstances (e.g. corrosion, unknown pressure rating of components, etc).

Whenever firewater pump tests are conducted, a procedure should be used that identifies the steps to be taken, readings to be recorded and safety precautions to be observed. The procedure should be reviewed and approved by the engineer in charge of the test operation. Generic procedures should not be used, instead procedures and diagrams are needed that identify specific valves to be opened and closed, equipment to be operated, and sequence of operation to be performed should be prepared for each installation. Only essential personnel should be present in the test area. Only personnel authorized to operate the equipment should be allowed to perform direct functions of startup and shutdown. Should the test engineer or operators notice unusual pump conditions, the pump should be shut down according to pre-arranged emergency shutdown procedures. Communication to all personnel involved in the test is the key to ensuring it is performed safety and effectively.

If at any time the firewater pump is placed out of service for performance testing, suitable equivalent backup support or arrangements should be provided. Equivalent protection may include assurance that secondary water supplies are available, such as a tank truck and pump or secondary stationary fire pumping units that are directly connected to the distribution system. Emergency instructions to immediately shut down the testing activities and return the pump under test to service may also be beneficial. Testing of firewater pumps should be coordinated with facility and operational departments.

13.3. FACTORY ACCEPTANCE TEST

All firewater pumps should be assembled and tested at the factory to ensure operability and compliance with purchase specifications. These tests should highlight any shortcomings in the design and manufacturing of the unit before it is placed in the field. The pump's certified flow performance curve should be prepared by the manufacturer from the results of these tests. The test should be performed against an agreed written procedure. Table 13-1 provides a list of required testing for the factory acceptance and witnessing of a firewater pump.

Lineshaft (i.e. borehole or deep-well pumps) cannot normally be tested at the factory with their complete length of column pipe due to a lack of a suitable well for such a test. Consequentially, the loss of head in the portions of the pipe that have been omitted from the test and the power absorbed by any shafting therein, cannot be measured. Any thrust bearing would also be more lightly loaded during this test than in the final site location. Should it be necessary to verify such data, it should be conducted at the final installation site for the system and agreement for it should be outlined in any purchase order for the unit. Figures 13-1 and 13-2 show the typical factory acceptance test arrangements for borehole pumps.

13.4. SITE ACCEPTANCE TEST AND COMMISSIONING

Each firewater pump should be subject to a full functional and flow performance test following its installation at the site to confirm proper installation. The manufacturer's certified shop test pump curve for the unit should be available for comparison during the site test. The site acceptance test (SAT) is usually witnessed or performed by the pump vendor. Insurance surveyors may also request to be present. Any discrepancies in the installation can be identified at this time and can be corrected by the vendor.

A formal written SAT should be prepared and used for the test. The test should include the verification of all instrumentation, control and alarm functions, startup of the pump by all designated modes, interface with other supervisory systems (i.e. Distributed Control System (DCS)), pressure and capacity verification and mechanical operability or completeness. If specialized tests or readings are required, e.g. vibrational baseline monitoring, motor voltages, baseline noise survey, etc. these should also be conducted at this time.

The startup and commissioning of a firewater pump for service should be included as part of the pre-startup safety review (PSSR) checklist for a new facility. Unless the installed firewater pump is provided to a high risk facility, alternative support

Table 13-1 Factory Acceptance Tests and Verifications

Hydrostatic Pressure Test.
Performance Test.
System Function Test.
Complete Unit Test.
Net Positive Suction Head Test.
Material Inspection Verification.
Dynamic Balancing of Impellers.
Dimensional Check of Purchaser's Interfaces.
Final Weight Check (offshore applications).
Drawing Approvals.
Vibrational Measurements (as required).
Noise Measurements (as required).

Figure 13-1 Factory Acceptance Testing

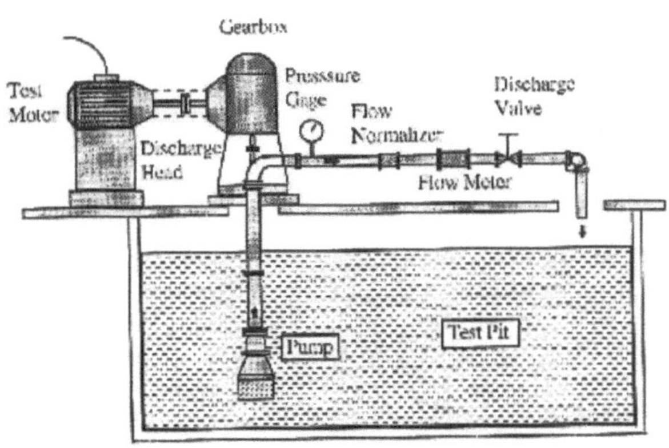

Figure 13-2 Factory Borehole Pump Testing Arrangements

measures should be provided or startup of the facility should be delayed until the firewater pumping system is operational.

13.5. PERIODIC PERFORMANCE TESTS

Since pumps will deteriorate due to wear from operation, corrosion or erosion, periodic performance tests should be performed to ensure the pump is capable of delivering the required amount of water and at an acceptable pressure. Rather than waiting until the pump fails or cannot meet a water delivery requirement, long term trend analysis is normally performed on the pump (i.e. the performance from year to year is plotted and an extrapolation of future deterioration is estimated). From these projections, the pump can be replaced before a detrimental lowering in performance occurs. Maintenance engineers have recognized the importance and savings achieved by this method. It is commonly called ''condition monitoring''. In some cases, these observations can indicate that a pump may last longer than the recommended manufacturer's suggestions and therefore may not have to be replaced as frequently, thereby producing a cost saving for the facility. In other cases, accelerated corrosion may be identified and corrective action taken to reduce the rate of corrosion, and the pump is replaced before it fails.

The performance of a firewater pump test is recognized by international standards and a list of some of the frequency and duration requirements are shown in Table 13-2. An individual company may also have its own periodic test requirements and frequencies. Pump performance testing should be recorded on standardized forms and copies maintained in the appropriate company files for later review by outside interested parties (i.e. insurance surveyors). Figures 13-3 and 13-4 provide examples of annual firewater pump test recording forms.

13.6. PUMP CURVE TEST POINTS

As general guidance, five test points for the firewater pump curve should be obtained. These should include: shutoff (0 percent), rated (100 percent) and overload (150 percent), with one point in between the shutoff and rated point i.e. at approximately 50 percent, and one in between the rated point and overload (125 percent). The reading taken at shutoff should be observed quickly or at a low flow, since at zero flow the fluid within the pump would have to absorb the entire power input and therefore heat up rapidly with injurious effects on the pump if allowed to continue.

13.7. FUEL EXAMINATION

Stored fuel for diesel engines should be sampled at the time of annual flow performance testing. The presence of micro-organisms, water or corrosion particles may indicate fuel contamination leading to reduced performance of the diesel engine. The date of the last engine fuel filter change and replacement schedule should be examined.

A fuel sample should be obtained from the fuel line feeding each engine. The samples should be taken at the time of the test and allowed to settle for 24 hours. Any suspended sediment or entrained water will separate. Water and particulates should collect at the bottom of the sample jar and a rough estimation can be made on the level of contamination.

Fuels that are contaminated will cause reduced performance of the diesel driver and therefore the output of the firewater pumps.

Table 13-2 Comparisons of Pump Test Requirements

Standard	Objective	Frequency	Duration
API RP 610	Run smoothly at rated load and speed	Weekly	Bring unit to normal operating temperatures
API RP 610	Confirm adequate pump capacity	Monthly	Until determination of capacity performance
Factory Mutual	Observe any problems and auto start	Weekly	30 minutes
Factory Mutual	Flow Performance Verification	Annually	Obtain three test points on pump curve (30 minutes minimum)
IRI IM 14.2.1	Prove good working order	Weekly	Until good working order is demonstrated
NFPA 20	Certified shop test	Completion of Fabrication	Until proper performance to NFPA is verified
NFPA 20	Field acceptance test	Completion of Installation	Until proper performance to NFPA is verified, but not less than one hour.
NFPA 20	Verify driver smooth performance at rated speed	Weekly	30 minutes or longer to obtain normal running temperatures
NFPA 20	Verify operation of engine with controller	Weekly	Automatic start and operate engine for 30 minutes
NFPA 25	Flow and capacity verification	Annually	Flow condition

Some companies have also used internal real-time video observation of the internal surfaces of pump columns or distribution piping to confirm its condition or amount of corrosion during maintenance inspections or suspected leaks.

13.8. SPECIFIC SPEED VERIFICATION

In some cases, the speed of a pump driver is not the same speed of the firewater pump, due to differences in the rated speed of the units and the provision of a geared coupling.

Pump No.:____

Location:____

Date of Test:____

Test Conducted By:____

Tide: Meters or Feet_______________________________

Correction for GPM: Rated RPM/Pump RPM x GPM = GPM Correction Factor (GCF)

Correction for PSI: $(\text{Rated RPM})^2/(\text{Pump RPM})^2$ x PSI = PSI Correction Factor (PCF)

Test Point Correction (Each change in driver (engine) speed requires a new correction factor)

GPM_______ x GCF (   ) =
Corr.GPM
1_____2_____3_____4_____5_____

PSI_________x PCF (   ) =
Corr.PSI 1_____2_____3_____4_____5_____

Driver:

Performance: Smooth_________Rough_________

Temperature Low______Normal______High______

Fuel Level ¼ ½ ¾ Full

Fuel Sample Clean_____Dirty_____Water____Other________

Comments:____

____

____________________ (Note any unusual noise, smoke, mechanical condition)

Right Angle Drive: Smooth_____Rough_____
Comments____

____

Pump Performance

Flow % Above___________
% Below____________Rated Curve

Pressure % Above___________
% Below_____________Rated Curve

Comments____________________________________ ______________________

Figure 13-3 Fire Pump Testing Form

FIRE PUMP TEST REPORT

Location ____________ Date______________

Pump No. ___________ Mfg. _____________Date______

Driver Type___________Mfg.______________Date______

Rating_______________Engine Hours____________

Time: Start__________ Stop__________

RPM________________ Gear Ratio________________

Calibration Dates:

Gages_________Flow Meter_________PSV_________

PSV Setting: Bar_____PSI_____

RPM	Pump Pressure			Flowmeter		Corrected to Rated RPM		Percent Rated Capacity
	Dis-charge	Suction	Net	Reading	x (GPM)	Net Head	Flow	

PUMP CURVE:

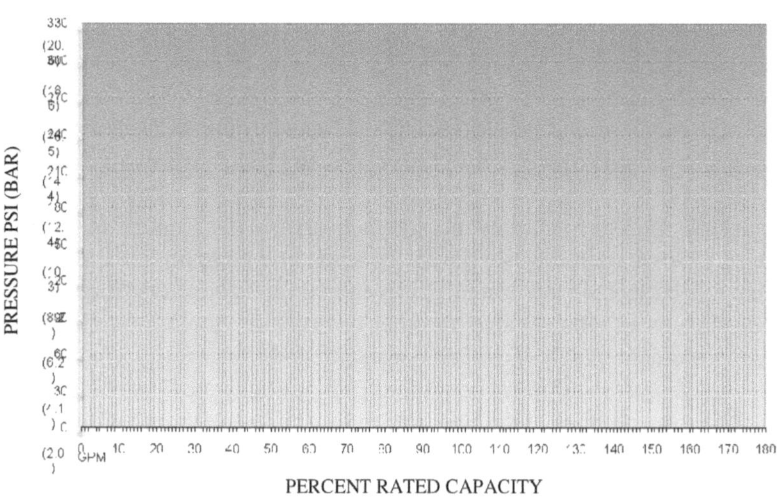

Figure 13-4 Supplemental Fire Pump Testing Form

Engine drivers for firewater pumps are normally provided with a tachometer which is used to verify the specific speed of the pump during flow testing against the factory test curve for a specific speed. American Petroleum Institute (API) Standard 610 recommends that the actual pump factory acceptance test be accomplished within three percent of the rated certified speed (i.e. revolutions) of the pump.

In theory, the characteristic pump curve assumes pump rotation at a constant specific rated speed. The revolutions of internal combustion engines are generally permitted to vary by up to 10 percent between shutoff and maximum load. For steam turbine drivers, up to eight percent is allowed. The speed of an electric motor is almost always constant. If the power circuit is overloaded, the speed of a motor may be reduced.

During testing of the firewater pump, the speed of rotation of the pump shaft or the driver should be independently confirmed through the use of revolution counter, calibrated tachometers or stroboscopic devices (i.e. commonly hand held strobe lights with digital speed readouts). If the speed is found to be different from the rated speed of the unit, the driver should be adjusted to as close as possible to the rated speed. Speed correction factors can be applied to determine the performance curve for the pump through affinity laws.

The following formulas are commonly employed:

$$\text{Flow at rated rpm} = (\text{Rated rpm}/\text{Observed rpm}) \times (\text{Observed Flow})$$
$$\text{Net Pressure at rated rpm} = (\text{Rated rpm}/\text{Observed rpm})^2 \times (\text{Observed Pressure})$$

13.9. ACCURACY OF TEST GAGES

The driver panel tachometer may not be an accurate enough instrument to record driver rotational speeds during a flow performance test of the firewater pump. Additionally, water pressure gages need to be periodically calibrated to ensure accuracy. The testing date of all gages should be recorded on test reports and the label provided with the gage itself.

Test gages for pressure and flow measurement are usually required to be located on a section of pipe where the water flow is constant and straight. For this reason they usually have to be located a minimum of five to 10 pipe diameters from any point of turbulence such as an elbow, valve, or other obstruction. Some multiple fitting may require the placement of a flow meter a considerable distance away in a straight piece of pipe to smooth out the turbulent conditions if highly accurate flow readings are to be obtained. The manufacturer of the flow measurement device should provide guidance in the allowable distances required.

13.10. WEEKLY TESTING

Fire pumps should be run weekly to verify their startup capability and general condition. Controllers are commercially available which provide automatic weekly startup

Weekly Fire Pump Test Sample Form

Test all fire pumps weekly. Enter correct settings in shaded column. Make sure all test results are within normal limits. If you find that repairs are needed, make them immediately and follow manufacturer's instructions.

Pump manufacturer			**Year installed**				
Manufacturer's model no.			**gpm/psi rating gpm ☐**		**psi ☐**		**rpm ☐**
District Office: Phone No. ______ Fax No. ______	Pump on ___ psi/bar/kPa		Jockey pump on ___ psi/bar/kPa		Pump off ___ psi/bar/kPa		Jockey pump off ___ psi/bar/kPa
Date tested							
By whom							
Pressure at pump startup method of start ______							
Motor running time (min)							
Suction pressure							
Discharge pressure							
Temp and tightness of stuffing box glands							
Level of water supplies suction tanks should be overflowed							
Temperature of water							
Pump room temperature							
Engine instrument readings							
RPM							
Oil pressure							
Temperature							
Last oil change ______ Next oil change ______							
Amps							
Fuel tank level should be at least three-fourths full							
Condition of crank case oil							
Condition of battery charger last time battery charged ______							
Cooling system temperature							
Cooling system strainer condition							

Annual pump flow test results satisfactory ____ Yes ____ No

Explain findings:

Provide a work order for immediate repair.

Follow impairment procedures (FM Item No. P9006).

Keep records on file for review by appropriate personnel.

Sign off when pump is restored to automatic: ________________

P8217A ed. 9/96

Figure 13-5 Weekly Fire Pump Test Sample Form – Page 1
Source: Courtesy of Factory Mutual Research Corporation, reprinted with permission.

of the firewater pump to meet this requirement. Insurance underwriters also have standard weekly test forms available (see Figures 13-5 and 13-6).

13.11. CONTROLLER AND INTERFACE TESTING

Confirmation of the fire pump controller operation capability should be performed periodically. Local and remote control points (i.e. control room) for firewater pump startup should be activated. Water should be bled from a pressure switch to simulate real fire conditions (i.e. a drop in firemain pressure due to firewater use). External connections to the controller circuitry should also be verified (i.e. fire detection and

Weekly Tests	
Automatic Start	Test the automatic start by opening a test line to reduce the system pressure.
Combustion Drivers	Run the engine at rated speed for at least 30 minutes while the pump discharges water through the circulation relief valve to an open drain. It takes at least 30 minutes to lubricate the engine and see what problems develop, if any.
Electric Drivers	Check operation of the starting devices and allow the pump to run at least five minutes.
Pressure Relief	Verify that relief valves operate properly. Usually this means water should not be discharging through them. Relief valves are designed to prevent pressure from exceeding the systems's design pressure.
Waterflow	Check waterflow to packing in the stuffing box. Slight leakage usually means water lubrication is adequate. Check the pump temperature. Feel the pump casing and bearings for overheating and signs of excess friction.
Water Supplies	Overflow the suction tanks or check them visually. In cold weather, make sure heat is provided to the supply lines and suction source. For open bodies of water, check the suction intake for possible obstructions. Verify that drought or dry conditions have not significantly reduced the water supply. On public supply booster pumps, check all valves on the suction line.
Pump Room Temp	It should be a minimum of 40°F (4°C). For internal combustion engines the recommended minimum temperature is 70°F (21°C); provide an engine jacket water heater set at 120°F (49°C); provide a lubricating oil heater if room temperature is not maintained above 70°F (21°C).
Diesel Engines	Make sure the engine is clean, dry and smooth running. Check: • Fuel tank levels: when it is three-quarters full, refill the tank. • Quality and quantity of the crankcase oil: replace oil if it has become fouled or has lost viscosity. • Battery charger and batteries: are they operating properly? • Cleanliness of the strainer in the water cooling system: clean the strainer when necessary; note temperature of the cooling system. • Proper operation of engine instruments: rpm, oil pressure, temperature, amps.
Monthly	check the specific gravity of the battery electrolyte.
Semiannually	examine the oil filter for sludge or foreign particles and insert a new filter cartridge. Have a diesel mechanic do a performance test.
Annually	do a waterflow test plus all tests listed above in the weekly, monthly and semi-annual tests. Your FM consultant can do the waterflow test. Record waterflow measurements and suction and discharge pressure readings for several different flow volumes. You can plot these on a rating chart and evaluate pump performance from year to year. The performance of any given test can be compared with the pump acceptance test, other yearly tests, and the pump manufacturer's characteristic pump curve.
Change	engine oil every year regardless of operation.
Take Flow and Pressure	readings for at least three well-spaced points on the pump curve (at churn or no-flow, near the pump rating, and at 150 percent flow point). If required flows cannot be met, or if there is more than a slightly noticeable change in performance, find the problem and correct it without delay. Flow water through a hydrant or hose header, or through a Factory Mutual Research Corporation Approved flow meter discharging to a safe location. Check water discharge rate from the heat exchanger and exhaust manifold. Compare the cooling water discharge rate with previous observations and the recommended rate.
Troubleshoot Problems	Identify deterioration in pump performance and correct problems before the pump becomes incapacitated. A reduction in capacity above ten percent requires investigation and possible repairs.

Figure 13-6 Weekly Fire Pump Test Sample Form – Page 2
Source: Courtesy of Factory Mutual Research Corporation, reprinted with permission.

alarm, deluge valve operations). Annunciation of remote alarms from the fire pump to plant utility monitoring systems (i.e. DCS) should also be confirmed.

13.12. FOAM PUMP TESTING

Foam pumps can deteriorate with age, use and corrosion just as firewater pumps. As foam application is critical in most instances, a method to verify the performance of foam pumps should be incorporated into the design of the foam system where use of these pumps is deemed critical. Failure of the foam pumping system may be initially indicated by a lower level of foam concentrate proportioning into the system, due to lower input pressures. A common test method is to incorporate a re-circulation bypass line from the foam pump discharge with a return into the concentrate storage tank. A flow meter and pressure gages are fitted as appropriate.

13.13. BASIC TEST PROCEDURE

The following is a generic test procedure that may be used to test the flow performance of fixed firewater pumps. The purpose of this procedure is to provide operations and engineering personnel with the basic steps and engineering knowledge to adequately and efficiently perform the performance testing that may be necessary for company policies and procedures. The procedure should be revised and modified as appropriate to highlight the exact items and valve set-ups that are necessary. The procedure should be performed according to local work permit arrangements and procedures.

Testing periods should be arranged with the appropriate management of any installation through approved work orders or permits for the facility. Careful observation of the equipment under test should be made to detect any abnormalities and signs of impending failures. The operational testing area should be restricted to the personnel solely involved in the testing. The procedure is as follows:

1. Obtain the manufacturer's pump curve for the unit to be tested. Confirm that the pump to be tested is properly identified, i.e. verify nameplate serial and equipment numbers, etc. Confirm driver rated revolutions per minute (rpm), and if fitted, the gearbox input and output rpm ratings.
2. Review results of the previous pump performance test. Note date of previous test, abnormalities recorded, actions taken to correct abnormalities, percentage below rated pump curve during previous test, etc.
3. Ensure that calibrated test gages, 0-1,380 kPa (0-200 psig), are installed in the suction and discharge piping of each firewater pump to be tested and are of sufficient pressure range. Record the date of calibration of test instruments on the performance data sheet. For vertical turbine pumps, a calculation of vertical head loss to the point of pressure reading is required, taking into account tide levels and seawater density.
4. Determine the flow measurement method to be used during the test, i.e. installed flow meter or manual pitot tube trough test outlets or the nearest available water outlets on the system via hoses and nozzles. If flow from outlets is to be used, ensure the area is restricted, the hoses are adequately secured, spray from nozzles will not impact personnel or property or disrupt ground surface and water runoff can be adequately disposed of. Ensure the flow measurement devices are calibrated and capable for the maximum flow output of the firewater pump.
5. Ensure independent measurement devices for verification of the driver (i.e. motor or engine) and pump shaft revolutions are available and accurate, e.g. strobe light hand held tachometer.
6. Ensure the pump recycle valve is closed if water is to be measured at the local water outlet or the system discharge is to be recycled into storage.
7. If a relief valve is fitted, verify piping is rated for maximum pump pressure output, then isolate relief valve only during testing period, if not, leave relief valve in service during testing, noting impact on test curve in report.

8. Ensure all personnel have been briefed on the hazards of the operation, safety requirements and emergency shutdown procedures. Ensure approved procedures are in use and work permits have been issued, as required.
9. Observe the condition of the fire pump controller panel. Verify no trouble indications, normal readings, and in operation.
10. Open the pump discharge valve approximately 50 percent.
11. Start up the pump to be tested (either through manual means or preferably by test of the automatic startup by low pressure or other means) and let it operate for a minimum of 15 minutes, for stabilization of the mechanical systems, warm up of fluids and verification of instrumentation operability.
12. Adjust the driver (i.e. engine) rpm to operate the pump as close as possible to its rated rpm.
13. Record five pressure (inlet and outlet) and flow readings for the pump near the following flow points: 0 percent (or as close to 0 without activation of the relief valve if necessary), 50 percent, 100 percent, 125 percent and 150 percent, simultaneously recording the rpm of either the pump or driver. The time of flow for each test point is dependent on the adequate stabilization of flow for an accurate instrumentation reading. Flow variance to be obtained by the opening or closing of water outlets from the pump discharge (i.e. additional test header outlets opened or isolation valve to flow meter opened).
14. Plot the test points against the rated pump curve, adjusted for the rated rpm of the unit, if required. If conditions permit, data should be plotted immediately during the test to indicate any obvious abnormality that may be corrected during the test, e.g. partially opened or closed valves, system components plugged, etc. Flow meters have a direct readout of flow, while pitot tube measurements require the reference to a hydraulic table for flow through circular outlets. A waterproof table of flow through the specific outlets to be flowed against with a pitot tube in advance, so efficient determination of flow levels can be made at the time of test (pump test data can be also inserted into a portable laptop PC that can immediately process the data in a flow analysis program).
15. Continuous monitoring of the driver and the pump should be maintained during the test. Specific attention should be given to the engine gages, gearing, cooling connections (especially cooling hoses), expected points of oil or fuel leakages and abnormal vibration or noises. Any unusual readings or observations should be brought to the attention of the maintenance personnel immediately.
16. Upon completion of the flow test, the pump should be shut down normally and the system returned to normal.
17. A final test report on the performance and condition of the firewater pump system tested (with recommendations, if required) should be prepared and submitted to management.

13.14. SUPPLEMENTAL CHECKS

For engine driven units, a sample of the fuel supply in the day tank should be taken. It should be analyzed for indications of water or sediment contamination. The sample should be allowed to stabilize for 24 hours to determine the contaminant content. Entrained water will collect at the bottom of the sample container and hydrocarbons fluids will collect on top of it. Particulates will settle to the bottom. The following checks should be carried out:

1. Certify that there are no leakages of oil or water for the engine driven units. The flexible connections for cooling water and fuel supplies should be checked for deterioration, cracks, etc.

2. Verify firewater pump startup on low pressure indication if such capability has been provided and tested while flow performance testing undertaken.
3. Verification of fire pump startup and flow and pressure indications in the plant utility monitoring system, e.g. DCS, should be confirmed if such indications are provided as part of the plant monitoring system. Remote stopping of the firewater pump should not be allowed.

13.15. CORRECTION FACTORS FOR OBSERVED TEST RPM TO RATED RPM OF DRIVER

$$\text{Flow at Rated RPM} = (\text{rated RPM}/\text{observed RPM}) \times (\text{observed Flow Rate})$$
$$\text{Net Pressure at Rated RPM} = (\text{rated RPM}/\text{observed RPM})^2 \times (\text{Net Pressure})$$

13.16. ADDITIONAL TESTING AND MAINTENANCE ASPECTS

Normal fire protection practices and standards recommend that fire pumps be tested annually to determine performance levels. Common practice in the process industries is to trend the flow performance to prepare predictive maintenance and replacement forecasts. Such forecasts can predict poor pump performance and prepare maintenance organizations to implement corrective actions before this occurs.

Fire pumps should be considered as critical to plant operations as process production pumps. Their operation and maintenance categories classified in plant schedules should be equal to or higher than all other equipment in the plant.

Operational and maintenance data files and field logbooks should be maintained and should record and document resolution of all testing outcomes, trouble indications, failures, maintenance performed, etc. Both internal and external auditors will require such documentation to be readily available and up-to-date. Failure to maintain such information will indicate a less than adequate management of fire water pumping systems by plant personnel.

14 HUMAN FACTORS AND QUALITY CONTROL

14.1. HUMAN FACTORS

Although most people have good intentions for performing work tasks, individuals are not machines and therefore there is some probability that human beings will forget, make mistakes or cause errors from time to time for various reasons. Human factors engineering tries to overcome these inherent human deficiencies and provide remedies. Consequentially, equipment needs not only to be designed clearly and simply, but also has to be fail-safe and foolproof.

14.2. IDENTIFICATION

Fire pumps are normally provided with a nameplate indicating its manufacture, capacity ratings, and serial number of the unit. A sign should be provided in the vicinity of a firewater pump, indicating the minimum pressure and flow required at the pump discharge flange to meet the system demand.

Pumps that can be started automatically or remotely should be provided with a warning sign stating the possibility of unattended driver and pump startup.

Nameplates should be affixed to all major components of the firewater pumping system, pump, driver, gearbox, etc. The nameplates should indicate the specific specifications for the unit, e.g. rated performance, capacity, etc. See Figure 14-1.

Pump rooms or pump houses should be labeled on the outside so they can be identified at a distance. This aids in emergency response actions.

The National Fire Protection Association (NFPA) 79, *Electrical Standard for Industrial Machinery*, can be referred to for highlighting and identifying equipment.

14.2.1. PAINTING

It is common practice within the industry to paint firewater equipment red. Most exposed surfaces of firewater pumps, and the piping to and from them are commonly painted or color banded red as an aid to locating these devices in an emergency. Painting also helps prevent corrosion activity. The American National Standards Institute (ANSI) A13.1, ANSI Z535.1, British Standard (BS) 1710 or the International Organization for Standardization (ISO) 3864: 1984, *Safety Colors and Safety Signs* may be referred to, to identify the exact color-coding required.

Fire Fighting Pumping Systems at Industrial Facilities. DOI: 10.1016/B978-1-4377-4471-2.00014-1

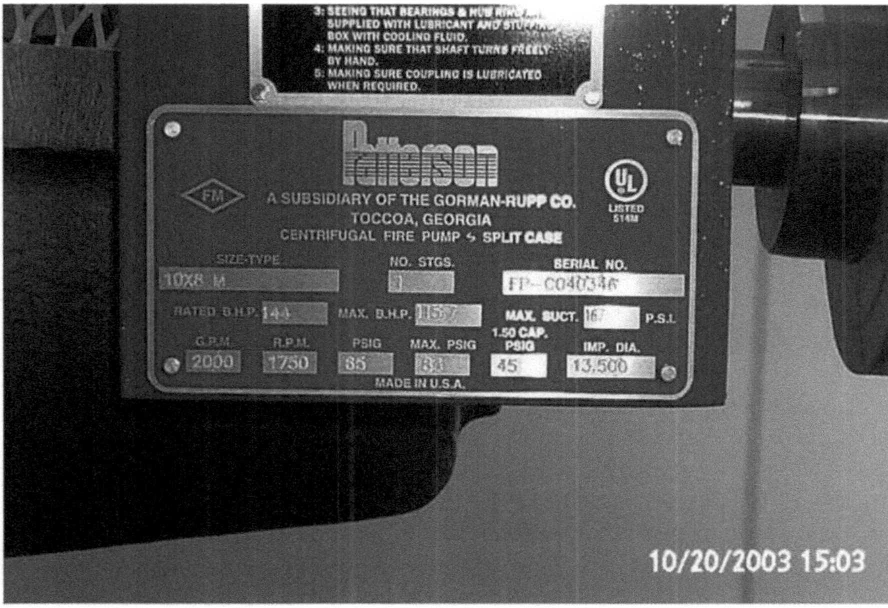

Figure 14-1 Fire Pump Nameplate

14.2.2. FLOW ARROWS

The direction of water flow to and from pumps should be clearly indicated on the piping arrangement. The arrows should be of a contrasting color to the piping and are normally white on red. They should be readily visible on the approach to the unit, or from the normal operator standby position. The casings of pumps are required to have a cast arrow indicating the direction of rotational flow.

14.3. STARTING INSTRUCTIONS

In some instances, there may be a need to start-up a firewater pump manually in an emergency, by personnel who are unfamiliar with the equipment. Without instructions in the proper method to start a firewater pump, personnel may be unable to start the unit, inadvertently inhibit the unit from starting, or damage the unit so that it is prevented from starting.

Manual starting instructions for all firewater pumps should be provided locally to the pump. These instructions may be as simple as indicating a switch to push, or as complicated as necessary, so long as they are plain, simple and foolproof. The instructions should be in the language(s) that is dictated for use at the facility. Instructions for starting the pump should be provided by the vendor, supplemented with any accessory means the owner has provided for supplemental starting means. These instructions should also be provided in the documentation maintained for the unit.

14.4. ACCESS

The pump driver controls should be easily accessible from at least two directions. This reduces the possibility that access to the unit will be blocked in an emergency, should manual startup or shutdown be required. Maintenance access for serviceable components should be provided. Additionally, removal and replacement of the firewater pump components should be considered. Access doors, hoisting capability and component removal clearances have to be evaluated.

Consideration should also be given to access for fire-fighting operations at the pump driver itself, in case it catches fire.

14.5. GUARDS

Protective guards should be provided to prevent accidental contact with rotating equipment, i.e. shafts, gear housings, couplings, etc. (see ANSI B15.1, *Safety Standard for Mechanical Power Transmission Apparatus*) and for hot surfaces (engine surfaces, exhaust manifolds and piping, etc.) by operators and individuals (see Figure 14-2). The guards for rotating equipment should be suitably highlighted so they are not inadvertently removed. Personnel should not be exposed to surface temperatures greater than 65 °C (149 °F) that are within easy reach.

Enclosures are also provided on fire pump controllers to prevent unauthorized changes to the setting of the system.

Figure 14-2 Fire Pump Coupling Guard

Suitable hazard warning signs should also be provided, in addition to guards, at locations where individuals may be subject to injury, e.g. where high voltages are used for electrical motor drives, electrical high voltage signs should be present.

14.6. NOISE LEVELS

When electrical motors for pumps are used indoors, the noise level generated may be excessive. This is usually a concern at high rotational speed, i.e. 3,600 rpm. Diesel engines are inherently noisy due to the combustion process. A sound level profile should be requested from the engine manufacturer indicating the various decibel levels across a frequency spectrum audible to the human ear (i.e. 125 hertz to an upper frequency of 8,000 or 10,000 hertz, especially between 500 and 4,000 hertz, the region most important for speech recognition).

Noise limitation is required to ensure verbal communications are understood, outside communication devices are heard (i.e. paging, sirens, etc.) and for human health and comfort. Where noise levels are not tolerable, an acoustical enclosure should be provided for the pump and driver assembly. The acoustical assembly should not hinder access or observation of the pump driver or its instruments and controls. When ventilation around the driver is required, the acoustical enclosure should not hinder the required airflow. The American Petroleum Institute (API) Standard 615, *Sound Control of Mechanical Equipment for Refinery Services*, should be consulted for the method of sound attenuation to be used.

If acoustical enclosures are not practical, hearing protectors (earmuffs or plugs) can be required in the area of high noise. Commercially available earmuff-type hearing protectors can decrease the sound level from about 10 decibels at 100 hertz to over 30 decibels for frequencies above 1,000 hertz.

14.7. EMERGENCY AND PRE-FIRE PLANS

Emergency and "pre-fire" plans should be prepared for an installation that intends to provide firewater protection for its facilities or processes. These plans should indicate the features of fixed firewater pumping systems, manning requirements at firewater pumps for emergencies, contingencies for pump failures or firemain ruptures, and backup support from other mobile firewater pumping equipment and alternative water supply sources.

Emergency and pre-fire plans should be endorsed by the appropriate company management and periodically reviewed and updated when changes occur.

14.8. DOCUMENTATION

Documentation of the design and installation of a firewater pump is vital to ensure that it is properly provided and maintained. Information on testing, spare parts, maintenance and service are also required.

Common documentation requirements are listed below:

- Piping and Installation Drawings (P & IDs).
- General Arrangement Drawings.

- Controller Wiring Diagram (with interconnections).
- Cause and Effects Chart (with set points).
- Pump Piping Isometrics.
- Data Sheets (Driver, Pump, Gearbox, Factory Certified Pump Curve, Noise Spectrum, accessories, etc.).
- Removal and Installation Sequence Instructions & Diagrams (for submerged vertical turbine pumps).
- Commission Spares, Spare Parts and Special Tools List.
- General Description & Operation Manuals for the pump, drive and ancillary devices or components.
- Maintenance and Lubrication Manuals (complete with requirements and recommended frequencies).
- Weight Control or Center of Gravity Diagrams (for offshore installations and major lifts).
- Corrosion Protection Measures.

Most of this documentation is normally supplied as part of the vendor package and the purchaser should indicate the format, style, the number of copies required, and whether company approval is required of the documents and, if so, the timing of the approvals during the equipment fabrication dates. Table 14-1 provides an example of a Documentation Submittals and Approval Form of the type used for offshore firewater pump procurement, where the quantity and time of submittal is indicated in the procurement request. The production of adequate documentation ensures quality control procedures are being adhered to. The assembled materials should be kept readily at hand in the maintenance or engineering libraries for the facility.

14.9. TRAINING

Facility personnel require training in the use of every item of equipment that is available or requires their operational support. Personnel cannot be expected to operate equipment in an emergency if adequate instruction or training in its use has not been given beforehand. Operators should be trained in the method of manually starting firewater pumps in case the emergency automatic start-up means is incapacitated.

14.10. SECURITY

Due to the fact that firewater pumps provide critical support to a facility, their control mechanisms must be secure in order to assure reliable operation. For example, the pumps should not be arbitrarily switched to manual when they should be set for an automatic start-up mode, as supply or discharge valves can be inadvertently or deliberately closed. NFPA 20 recommends that controller cabinets are locked and valves are supervised. Operators should periodically visit the firewater pump location to confirm its operating condition.

Table 14-1 Document Submittals and Approvals

Documentation	For Approval		Final Data Package	
	Quantity	**Within Days**	**Quantity**	**Within Days**
General:				
Descriptive literature				
Typical drawing				
Manufacturing guarantee				
Technical manual				
Data:				
Completed data sheet				
Testing Procedure/configuration				
Performance curve				
Calculation sheet				
Torsional vibration Analysis				
Drawings:				
Overall dimensional				
General arrangement				
Impeller				
Bowl assembly				
Strainer				
Shafting/coupling				
Hypochlorite sys. (if required)				
Sub-vendors Drawings				
Assembly diagram				
Weld procedure				
Hydrotests				
Pump performance Test				
Schedules:				
Bill of materials				
Parts list/drawing list				
Recommended spares (2 years)				
Commissioning spares				
Manuals:				
Installation instructions				
Commissioning instructions				
Operation instructions				
Maintenance instructions				
Inspection/certifications:				
As required by codes/company				
Construction materials:				
Material traceability certificates				
Inspection certificates				
Inspection records				
Radiograph reports				
Radiographs				
Test certification				
Hydrotests				
Pump performance test				

14.11. QUALITY CONTROL

To ensure an effective firewater pumping system is provided for a facility, quality control (QC) measures must be provided throughout the design, manufacturing, testing, shipment, installation and commissioning of the unit.

Commonly, ISO 9000 series standards are referenced to demonstrate adequate quality assurance (QA) applications. The ISO 9000 series for contractual applications generally indicates the following:

- ISO 9001—QA in Design, Development, Production, Installation and Servicing.
- ISO 9002—QA in Production and Installation.
- ISO 9003—QA in Final Inspection and Testing.

In general, a quality control plan from a vendor should be provided to the purchaser of a firewater pump for review. It should contain the following information:

- Designation of responsibilities for QC on the scope of work.
- Activities to be performed, i.e. all stages which effect or measure the quality of the projects.
- All internal inspections and tests, contract review, process stages, procedures, and operative qualifications.
- Independent or purchaser inspections and tests.
- Procedures to be used to control performance and ensure compliance with the purchase order (i.e. design controls, inspections, tests, material traceability, quality audits, document controls, work instructions, method of corrective actions or repairs, etc.).
- Acceptance criteria required by the contract.
- Verifying documents to be provided.
- Hold and witness points for the purchaser or the purchaser's agent.

Satisfactory inclusion of these items in the plan and performance during the contract will ensure that a firewater pump meets specifications and performance.

APPENDICES

Fire Fighting Pumping Systems at Industrial Facilities. DOI: 10.1016/B978-1-4377-4471-2.00015-3

A.1 SELECTED MAJOR INCIDENTS AFFECTING THE PERFORMANCE OF FIREWATER PUMPING SYSTEMS

The following is a brief listing of several major incidents occurring in the processing industries which have directly affected the operation of the facility firewater pumps. This information is gathered from published public sources describing events which occurred during the incidents. Most of these incidents highlight the severely impacted firewater system as a result of primary or secondary effects from vapor cloud explosions. It is interesting to note that the average loss for these incidents is approximately $233 million. If the average cost to install a firewater system at an industrial facility is roughly $500,000; the average pumping system cost to average financial impact is approximately 0.2 percent.

August 21, 1991
Chemical Storage Facility, Melbourne, Australia
Lighting strike caused an explosion and fire. Fixed fire suppression systems were damaged in the initial explosion, resulting in the total reliance on manual fire fighting to extinguish the fire. Business interruption loss was estimated at $40 million.

March 12, 1991
Ethylene Oxide Manufacturing Unit, Seadrift, Texas
Explosion at plant caused loss of all utilities, and impacted several water sprays causing inadvertent operation. The operation of these systems caused a substantial loss of available water for manual fire fighting. A business interruption loss of $60 million and a downtime of one year were estimated.

December 24, 1989
Refinery, Baton Rouge, Louisiana
A 203.2 mm (8-inch) pipeline ruptured, allowing a vapor cloud to form which subsequently ignited. The resultant blast resulted in the partial loss of electricity, steam and firewater. Lines for docks fire pumps were also damaged. Water for fire fighting was supplied by remaining firewater pumps and mobile fire trucks tanking draft from alternative sources. $44 million loss was estimated.

July 6, 1988
Offshore Platform, North Sea
Explosion and subsequent fires destroyed the entire offshore oil and gas production platform. The initial explosion was thought to have caused a firewall to impact facility firewater pumps and distribution firewater mains. Estimated losses were at $1 billion.

May 5, 1988
Refinery, Norco, Louisiana
Pipeline corrosion caused a vapor cloud to occur, resulting in a massive explosion in the plant. The blast wave caused immediate loss of all utilities including firewater and four diesel firewater pumps. Loss was estimated at $300 million.

November 14, 1987
Petrochemical Plant, Pampa, Texas
Unconfined vapor cloud explosion at a large petrochemical plant. The blast severed sprinkler system piping and caused the underground firemain to rupture. The firehouse collapsed with fire apparatus inside. The fire was extinguished after 12 hours and resulted in $185 million in damages.

November 19, 1984
LPG Terminal, Mexico City, Mexico
A 203.2 mm (8-inch) line ruptured resulting in an explosion and fire. A series of five boiling liquid expanding vapor explosions (BLEVEs) occurred a short time later. The terminal's firewater system was impacted in the initial blast. Water was transported to the terminal by 100 tank cars for storage tank cooling purposes.

July 23, 1984
Refinery, Romeoville, Illinois
Failure of process column caused a vapor cloud to form. The vapor cloud ignited resulting in various fires and the occurrence of a BLEVE. Initial explosion disrupted all electric driver power at the refinery leaving a firewater pump inoperable and also shearing off a hydrant, causing a reduction in firewater pressure from the remaining firewater pumps in operation. Mobile firewater pumper trucks and a fireboat eventually provided water at sufficient pressures to conduct fire fighting. Losses were estimated at $143 million.

May 30, 1978
Refinery, Texas City, Texas
Tank overfilled causing vapor cloud to form. Resultant explosion caused additional tanks and vessels to rupture. Included in the impacts were the firewater storage tank and electric firewater pumps. Two diesel firewater pumps remained in service. $93 million dollar loss estimated.

A.2 FM GLOBAL REPORT ON FIRE PUMP LOSS HISTORY

INTRODUCTION

A study of fire pump failures over a 20-year period was reported in September 2010 by FM Global in its data sheet on fire pumps (No. 3-7). This study reported 73 losses with a total gross loss of $417 million (averaging to a $643,000 loss per incident). The largest loss was due to failure of utility power to the electric fire pump and resulted in a single $42 million loss. The losses incurred for the next largest incidents were $31 million, $29 million and $26 million and were the result of a diesel engine failure to start, and others for the fire pumps being "out of service" prior to the fire incident.

The losses highlight that a single fire pump failure may result in multi-million dollar losses and that these losses can be easily prevented by properly maintained fire pump operation capabilities.

SUMMARY OF CAUSE OF PUMP FAILURES BY MONETARY IMPACT ($)

- 51% impaired prior to fire
- 28% failed to start
- 15% shut-off prematurely
- 6% failed during the fire.

SUMMARY OF CAUSE OF PUMP FAILURES BY NUMBER OF LOSSES

- 38% failed to start
- 35% impaired prior to fire
- 18% failed during the fire
- 9% shut-off prematurely.

B PURCHASE DATA/SPECIFICATION SHEET

The essential data relating to any fire pump must include pressure and volume ratings. Secondary data concerns the rotation speed, power required and cost. Specification sheets are provided to prospective vendors to obtain preliminary quotes for pump supply and are certified after purchase agreements. They are maintained in the facility files as part of the equipment data documents.

Data Sheet Centrifugal Fire Pump	**Nolan Engineering Company**
Project No.:	Facility:
Equipment No.:	Number Required:
Operating Conditions	
Water Conditions:	Temperature Range (°C)
Fresh/Saltwater	Ambient Air:
Contaminates:	Water:
Corrosion Concerns:	Specific Gravity:
Vapor Pressure at P.T.:	Viscosity at P.T.:
Flow Requirements:	
Rated Flow at P.T.	Discharge Pressure:
Suction Pressure (max.):	Differential Pressure:
NPSH Available:	Engine Cooling Water Requirements:
Applicable Standard: NFPA 20/UL 448/BS EN ISO 9906:2000/API 610	Company Standards/Specifications:
Construction:	
Casing -Mounting:	Split:
Centerline/Foot/Bracket/Vertical	Axial/Radial
Type:	Tapped Openings:
Single Volute/Double Volute/Diffuser/Bowl	Vent/Drain/Gage Connections
Suction Nozzle:	Discharge Nozzle:
Size:	Size:
Rating:	Rating:
Facing:	Facing:
Position:	Position:
Impeller Diameter:	Impeller Type:
Rated:	
Minimum:	
Maximum:	

(*Cont.*)

Fire Fighting Pumping Systems at Industrial Facilities. DOI: 10.1016/B978-1-4377-4471-2.00030-X

Data Sheet Centrifugal Fire Pump	Nolan Engineering Company
Mfr's Bearing	Coupling Mfr.
No. Radial:	Model:
Thrust:	Size:
Packing:	Mech. Seal:
Manufacturer	Manufacturer:
Type	Model
Size No. of Rings	API Class Code
Gearbox:	General Arrangement:
Manufacturer	Skid Assembly
Model:	Baseplate
Service Factor:	Enclosure
Rotation	Foundation Bolts
Marine Package:	Discharge Head
Cooling:	

Materials	(ASTM No. or Other Spec.)
Casing:	
Bearing Bracket:	
Impeller:	
Impeller Wear Ring:	
Casing Wear Ring:	
Shaft:	
Shaft Sleeve:	
Gland/seal end plate:	
Gaskets:	
Drain/Vent Piping:	
Suction Strainer:	
Fasteners:	

Pump Drive

Motor Drive:	Engine Drive:
Manufacturer:	Manufacture:
Model	Model:
Volts/Phase/Cycles	Power Rating (kW or Hp):
RPM	RPM
Frame No.	
Enclosure (NEMA)	
Steam Drive	
RPM	
Steam Pressure Range:	

Performance:	(Completed by Vendor)
Proposal Curve No.:	Pump Model No.:
NPSH Required:	RPM:
No. of Stages:	BHP Max BHP rated Impeller:
Design Efficiency	Max. Rated Head:
Rotation Direction	Cooling Water Req.
Weight:	

BIBLIOGRAPHY

American Gear Manufacturers Association (AGMA) (1988). American National Standards Institute (ANSI)/AGMA Standard 2003-A86, *Rating the Pitting Resistance and Bending Strength of Generated Straight Bevel, Zerol Bevel, and Spiral Bevel Gear Teeth,* Arlington, VA, AGMA.

American Petroleum Institute (API) (2007). API RP 14C, *Recommended Practice for Analysis, Design, Installation and Testing of Basic Surface Safety Systems for Offshore Production Platforms,* 7th edn. Washington, DC, API.

American Petroleum Institute (API) (2007). API RP 14G, *Recommended Practice for Fire Prevention and Control on Open Type Offshore Production Platforms,* 4th edn. Washington, DC, API.

American Petroleum Institute (API) (2004). API Standard 610, *Centrifugal Pumps for Petroleum, Petrochemical and Natural Gas Industries,* 10th edn. Washington, DC, API.

American Petroleum Institute (API) (1987). API Standard 615, *Sound Control of Mechanical Equipment for Refinery Service,* Washington, DC, API.

American Petroleum Institute (API) (2009). API Standard 676, *Positive Displacement Pumps - Rotary,* 3rd edn. Washington, DC, API.

Association of Offshore Diving Contractors (AODC) (1991). *Protection of Water Intake Points for Diver Safety, AODC 055,* London, UK, AODC.

American Society for Testing of Materials (ASTM) (1988). ASTM F-1166-88, *Standard Practice for Human Engineering Design for Marine Systems, Equipment and Facilities*, ASTM.

American Water Works Association (AWWA) (2006). American National Standards Institute (ANSI)/AWWA E102-06, *Standard for Submersible Vertical Turbine Pumps,* 1st edn. Denver, CO, AWWA.

American Water Works Association (AWWA) (1989). *Distribution System Requirements for Fire Protection, M31*, Denver, CO, AWWA.

Benaroya, A. (1978). *Fundamentals and Applications of Centrifugal Pumps for the Practicing Engineer*, Tulsa, OK, Petroleum Publishing Company.

British Standard, BS EN ISO 9906:2000 (BSI) (2000). *Rotodynamic Pumps. Hydraulic Performance Acceptance Tests. Grades 1 and 2.* London, UK, BSI.

Casey, J.F. (1974). *Fire Service Hydraulics*, New York, Dun-Donnelley Publishing Corporation.

Clough, I. (ed) (2010). *The 100 Largest Losses 1972–2009, Large Property Damage Losses in the Hydrocarbon-Chemical Industries,* 21st edn. London, UK, Marsh Global Risk Engineering.

Coon, J.W. (1991). *Fire Protection, Design Criteria, Options, Selection*, Kingston, MA, R.S. Means, Co. Inc.

De Camp, L.S. (1990). *The Ancient Engineers*, New York, Dorsett Press.

Department of Energy/The Hon Lord Cullen (1990). *The Public Inquiry into the Piper Alpha Disaster*, London, UK, HMSO.

Des Norske Veritas (DNV) (2009). *OREDA Offshore Reliability Data Book*, 5th edn. Olso Norway, DNV.

Dickerson, C. (1995). *Pumping Manual*, 9th edn. London, UK, Elsevier Science.

EEMUA, EEMUA 107:1985 (1992). *Recommendations for the Protection of Diesel Engines for Use in Zone 2 Hazardous Areas.* EEMUA.

Fire Fighting Pumping Systems at Industrial Facilities. DOI: 10.1016/B978-1-4377-4471-2.00016-5

Energy Industries Council (EIC) (1992). *A General Specification for Firewater Pump Packages* (Draft), London, UK, EIC.
Factory Mutual Global, Property Loss Prevention Data Sheet 3–7 (2010). *Fire Protection Pumps*, Norwood, MA, Factory Mutual Global.
Factory Mutual (1976). *Handbook of Industrial Loss Prevention*, 2nd edn. New York, McGraw-Hill.
Factory Mutual Research (2007). *Approval Standard, Centrifugal Fire Pumps Split-Case Type, (Axial or Radial), Class Number 1311*, Norword MA, Factory Mutual Research Corporation.
Factory Mutual Research (1999). *Approval Standard, Centrifugal Fire Pumps (Vertical-Shaft, Turbine Type), Class Number 1312*, Norwood MA, Factory Mutual Research Corporation.
Factory Mutual Research (2008). *Approval Standard, Centrifugal Fire Pumps (Horizontal, End Suction Type), Class Number 1319*, Norwood MA, Factory Mutual Research Corporation.
Factory Mutual Research (2002). *Approval Standard, Controller for Electric Motor Driven and Diesel Engine Driven Fire Pumps, Class Number 1321/1323*, Norwood MA, Factory Mutual Research Corporation.
Factory Mutual Research (1975). *Approval Standard, Right Angle Gear drives, Class Number 1338*, Norwood, MA, Factory Mutual Research Corporation.
Hickey, H.E. (1980). *Hydraulics for Fire Protection*, Boston MA, National Fire Protection Association (NFPA).
Hydraulics Institute (1983). *Standards for Centrifugal, Rotary and Reciprocating Pumps*, 14th Edition. Cleveland, OH, Hydraulic Institute.
International Maritime Organization (IMO) (1986). *Safety of Life at Sea (SOLAS)*, London, UK, IMO.
Ingersoll Rand Company (1981). *Cameron Hydraulic Data*, 17th edn. New York, NJ, Ingersoll Rand.
Institute of Petroleum (2005). *Model Code of Safe Practice in the Petroleum Industry, Part 15, Area Classification for Petroleum Installations*, 3rd edn. London, Institute of Petroleum.
Insurance Services Office (ISO) (1974). *Guide for Determination of Required Fire Flow*, New York, NY, ISO.
International Standardization Organization (ISO) (1999). *9906, Rotodynamic Pumps – Hydraulic Performance Acceptance Tests,* Grades 1 and 2,, ISO.
Isman, K. E., and Puchovsky, M. T. (eds.). (1998). *Fire Pump Handbook.* Qunicy MA, National Fire Protection Association.
Jacobs, T.A. (1993). *The Fire Engine*, Greenwich, CT, Brompton Books Corp.
James, P., Thorpe, N. (1994). *Wonders of the Past! Ancient Inventions*, New York, NY, Ballantine Books.
Karassik, I.J., et al. (1986). *Pump Handbook*, 2nd edn. New York, NY, McGraw-Hill.
Lewis, J.R., Mercer, A.D. (1984). *Corrosion and Marine Growth on Offshore Structures*, Chichester, UK, Ellis Horwood Limited.
Loss Prevention Council (LPC) (2009). *LPC Rules for Automatic Sprinkler Installations*, Borehamwood, Hertfordshire, UK, LPC.
National Fire Protection Association (NFPA) (2008). *Fire Protection Handbook*, 20th edn. Quincy, MA, NFPA.
National Fire Protection Association (NFPA) (2010). NFPA 11, *Standard for Low Expansion Foam and Combined Agent Systems,* Quincy, MA, NFPA.
National Fire Protection Association (NFPA) (2010). NFPA 13, *Standard for the Installation of Sprinkler Systems,* Quincy, MA, NFPA.
National Fire Protection Association (NFPA) (2010). NFPA 14, *Standard for the Installation of Standpipe and Hose Systems,* Quincy, MA, NFPA.

National Fire Protection Association (NFPA) (2007). NFPA 15, *Standard for Water Spray Fixed Systems for Fire Protection,* Quincy, MA, NFPA.

National Fire Protection Association (NFPA) (2010). NFPA 20, *Standard for the Installation of Centrifugal Fire Pumps,* Quincy, MA, NFPA.

National Fire Protection Association (NFPA) (2010). NFPA 24, *Standard for the Installation of Private Fire Service Mains and Their Appurtenances,* Quincy, MA, NFPA.

National Fire Protection Association (NFPA) (2011). NFPA 25, *Standard for the Inspection, Testing, Maintenance of Water-based Fire Protection Systems,* Quincy, MA, NFPA.

National Fire Protection Association (NFPA) (2011). NFPA 31, *Standard for the installation of Oil Burning Equipment,* Quincy, MA, NFPA.

National Fire Protection Association (NFPA) (2010). NFPA 37, *Standard for the Installation and Use of Stationary Combustible Engines and Gas Turbines,* Quincy, MA, NFPA.

National Fire Protection Association (NFPA) (2011). NFPA 70, *National Electrical Code,* Quincy, MA, NFPA.

National Fire Protection Association (NFPA) (2007). NFPA 79, *Electrical Standard for Industrial Machinery,* Quincy, MA, NFPA.

National Fire Protection Association (NFPA) (2010). NFPA 110, *Standard for Emergency Power Supplies,* Quincy, MA, NFPA.

Nolan, D.P. (1995). Rare Corrosion Failure of a 90/10 Copper Nickel Firewater Pipe. NACE 95 Conference Proceedings, Paper # 95-265, NACE International.

Nolan, D.P. (1995). A Statistical Review of Fire and Explosion Incidents in the Gulf of Mexico, 1980-1990, *Journal of Fire Protection Engineering,* Society of Fire Protection Engineers. Vol.VII, No. 3.

Nolan, D.P. (2010). *Handbook of Fire and Explosion Protection Engineering for Oil, Gas, Chemical and Related Facilities*, 2nd edn. Park Ridge, NY, Elsevier Publications.

Oil Companies International Marine Forum (1974). *International Oil Tanker and Terminal Safety Guide*, 2nd edn. London, UK, Applied Science Publishers.

Rayner, R. (1995). *Pump Users Handbook*, 4th edn. Tulsa, OK, Penwell Publishing Company.

Strandh, S. (1979). *A History of the Machine*, New York, NY, A & W Publishers, Inc.

Society of Fire Protection Engineers (SFPE)/National Fire Protection Association (NFPA) (2008). *The SFPE Handbook of Fire Protection Engineering*, 4th edn. Boston, MA, NFPA 4-51 to 4-54.

U.K. Department of Energy (1978) S. I. 611. Offshore Installations: The Offshore Installations (Fire-fighting Equipment) Regulations 1978, London, UK, HMSO.

Underwriter's Laboratories (UL) (2009). UL-218, *Fire Pump Controllers,* 2nd edn. Northbrook, IL, UL.

Underwriter's Laboratories (UL) (2007). UL 448, *Standard for Centrifugal Stationary Pumps for Fire Protection Service,* 10th edn. Northbrook, IL, UL.

Underwriter's Laboratories (UL) (2008). UL 448A, *Standard for Flexible Couplings and Connecting Shafts for Fire Pumps,* 1st edn. Northbrook, IL, UL.

Underwriter's Laboratories (UL) (2009). UL 448C, *Standard for Stationary, Rotary-Type, Positivie Displacement Pumps for Fire Protection Service,* 1st edn. Northbrook, IL, UL.

Underwriter's Laboratories (UL) (2007). UL 1247, *Standard for Diesel Engines for Driving Stationary Fire Pumps,* 5th edn. Northbrook, IL, UL.

Underwriter's Laboratories (UL) (2004). UL 1478, *Standard for Fire Pump Relief Valves,* 4th edn. Northbrook, IL, UL.

Warring, R.H. (1984). *Pumping Manual,* 7th edn. Houston, TX, Butterworth-Heinemann.

Whitele, R. (1986). Engineering Solutions Control the Surge. Offshore Engineer.

XL Global Asset Protection Services (XL GAPS) (2007). GAP.14.0.1, *Fire Protection Water Supplies,* Stamford, CT, XL GAPS.

XL Global Asset Protection Services (XL GAPS) (2010). GAP.14.2.1, *Stationary Pumps for Fire Protection – NFPA 20-2010,* Stamford, CT, XL GAPS.

XL Global Asset Protection Services (XL GAPS) (2005). GAP.14.2.1.1, *Centrifugal Fire Pump Acceptance Tests,* Stamford, CT, XL GAPS .

XL Global Asset Protection Services (XL GAPS) (2005). GAP.14.2.1.3, *Testing of Fire Pump Controller Features,* Stamford, CT, XL GAPS.

Yedidiah, S. (1980). *Centrifugal Pump Problems, Causes and Cures,* Tulsa OK, Petroleum Publishing Company.

GLOSSARY

Absolute Pressure a unit of measurement that accounts for the atmospheric pressure available on the free surface of the fluid or the pressure existing in a closed reservoir.

Approved requirement of a code, standard, device or item of equipment that is duly sanctioned, endorsed, accredited, certified, listed, labeled or accepted by a nationally recognized authority or agency as satisfactory for use in a specified manner.

Aquifer an underground geological formation that contains a sufficient amount of saturated permeable material (e.g. limestone, sandstone) to yield water in significant quantities (i.e. able to support human needs).

Automatic Transfer Switch (ATS) an electrical switch which is self-actuating for transferring one or more load conductor connection(s) from one source of power to another.

Biocide chemical material used to control or inhibit the growth of marine organisms.

Booster Pump a pump that obtains suction from a public service water main or private-use water system (i.e. positive pressure water source) for the purposes of increasing the effective water pressure.

Break Tank a fire protection water tank that is of not sufficient size for the full fire protection demand, but uses automatic refill capability to achieve adequacy.

Cavitation formation of a partial vacuum (creating gas bubbles) in a liquid by a swiftly moving solid body (e.g. a propeller). The generation and collapse of the gas bubbles produces vibration and sometimes causes a mechanical strain on the pumping system reducing performance and causing deterioration of the pump, particularly its impeller.

Caisson length of pipe extending vertically downwards from offshore installation into the sea to a position below the lowest sea level. It is a means of protection for submerged firewater pumps and their supply columns. May also be referred to as a conductor or stilling tube.

Centrifugal proceeding or acting in a direction away from the center or axis.

Centrifugal Pump a pump that utilizes a rotating impeller to increase the pressure and flow of the fluid. The fluid enters the pump's enclosed impeller along or near the rotating axis and is accelerated by the impeller and is then discharged outward into a piping system.

Circulation Relief Valve a valve on a fire pump whose function is to discharge an amount of water to prevent overheating of the pump during shut-off conditions.

Classified Area an area or zone defined as a three-dimensional space in which a flammable atmosphere is or may be expected to be present in such frequencies as to require special precautions for the construction and use of electrical apparatus and hot surface exposures that can act as an ignition source.

Column assembly of vertically straight pipes from a submerged pump which direct water to a desired location.

Fire Fighting Pumping Systems at Industrial Facilities. DOI: 10.1016/B978-1-4377-4471-2.00017-7

Controller the cabinet, motor starter, circuit breaker, disconnect switch or other devices for starting and stopping electric motor and internal combustion engine-driven firewater pumps.

Corrosion physical change, usually deterioration or destruction that is caused through chemical or electrochemical interaction.

Critical Function an operation or activity which is essential to the continuing survival of a system. Any of those functions which are vital to the life of the system.

Diesel Engine Fire Pump Controller fire pump device to automatically control (e.g. initiate startup) the operation of a diesel engine driven fire pump based on input signals.

Distributed Control System (DCS) computer based control system that segments portions of the control to various locations.

Drawdown the vertical difference between the level of water while pumping and the level of water during static conditions. The pumping level is usually taken while at maximum capacity.

Driver the electric motor or diesel engine that imparts the rotation force to operate a fire pump.

Electric Fire Pump Controller fire pump device to automatically control (e.g. initiate startup) the operation of an electrically driven fire pump based on input signals.

End Suction Pump a pump having its suction nozzle on the opposite side of the casing from the stuffing box and having the face of the suction nozzle perpendicular to the longitudinal axis of the shaft.

Failsafe a system design or condition such that the failure of a component, subsystem or system or input to it, will automatically revert to a predetermined safe static condition or state of least critical consequence.

Fault Tree Analysis (FTA) a method for representing the logical combinations of various system states that lead to a particular outcome.

Fire Flow a common term in the fire protection profession for the required fire water delivery rate for a particular occupancy.

Fire Pump a pump specifically designed and dedicated to deliver a specified rate of water flow at a specified pressure to a fire protection system.

Fire Pump Controller a device that is arranged to operate in a specified manner the starting and stopping of a fire pump driver, as well as monitor and signal the status and condition of the fire pump package.

Fire Pump Package an assembled unit of fire pumping capability that consists of a fire pump, driver, controller and accessories.

Firewater Pumping System system of equipment that imparts momentum to water supplies in a contained pipe network in adequate pressures and volumes to support fire protection activities.

Firewater System a water delivery system consisting of a water supply and distribution network to provide firefighting water for the control and suppression of fire incidents and exposure cooling requirements.

Flexible Coupling a device used to connect a driver to a pump which usually can compensate for small misalignments and dampen vibration.

Flow Meter a device for measuring the quantity of flow through a given area.

Foam Pump Controller special electric or diesel engine fire pump controllers utilized for the management of foam concentrate fire protection pumps.

Foolproof so plain, simple, obvious and reliable as to leave no opportunity for error, misuse or failure to implement the correct action.

Highly Protected Risk (HPR) term used within the insurance industry to describe a property risk that has sprinkler protection and is considered a superior facility from a fire protection viewpoint (i.e., low probability of loss), therefore it has a very low insurance rate compare to other industrial risks.

Hollow Shaft a shaft with the capability to accept the solid shaft of a pump and is commonly found on right angle gear drives and electric motors. It allows the adjustment of pump impellers within a bowl assembly and the installation of a non-reverse ratchet in an electric motor or gear drive.

Horizontal Shaft Centrifugal Pump a centrifugal type pump characterized by its shaft being affixed or mounted to the impeller(s) in a horizontal plane.

Horizontal Split-Case Pump a centrifugal type pump characterized by a housing that is split parallel to the shaft.

In-Line Pump a centrifugal pump whose drive unit is supported by the pump which has its suction and discharge flanges on the same centerline.

Jockey Pump a pump that maintains constant pressure on system piping and is sometimes referred to as a pressure maintenance pump. Jockey pumps do not meet the same levels of approval, integrity, and reliability as required for main firewater pumps.

Lineshaft shaft used to transmit power from a driver to a pump shaft. Commonly referred to in the description for vertical turbine type pumps.

Listed tested and approved by a recognized independent evaluation organization to a recognized standard.

Mass Elastic Torsional Analysis an engineering review and evaluation of torsional forces and linear resonant frequencies which may occur from the operating speed range of rotating equipment to determine if they can be eliminated to prevent damage.

Maximum Working Pressure the highest pressure developed at the pump discharge flange under any anticipated condition of suction pressure and pump flow.

Mechanical Seals a device to form a seal between a pump shaft and stationary components. It consists of a primary seal that is achieved by two flat, lapped faces placed perpendicular to the pump shaft. The contact between the two flat surfaces prevents leakages. One is held stationary in a housing and the other is fixed to the rotating shaft. Dissimilar materials are used for the two seals to prevent adhesion of the faces. They can be used in place of compression (soft) seals.

Multiple Stage Pumps pumps that contain more than one impeller on the same shaft. The number of stages is determined by the number of impellers.

Net Positive Suction Head (NPSH) is the suction head absolute available less the sum of the suction system and component friction losses, the static height the fluid must be lifted below the centerline of the pump impeller (on static lift applications), and the lessening absolute pressure when the suction vessel is under a vacuum.

Net Positive Suction Head Available (NPSHa) is similar to the NPSH except that the vapor pressure of the fluid at pumping temperature has been deducted.

Net Positive Suction Head Required (NPSHr) is the characteristic of a pump and is normally shown on the pump's performance curve.

Orifice Meter an instrument which measures the flow through a pipe by the use of the difference in pressure on the upstream and downstream sides of an orifice plate.

Overload term used to describe a fire pump that is discharging more water than it is designed to discharge.

Pitot Tube as employed in fire protection, an instrument for measurement of water velocity pressure. By relationships of water velocity pressure to the size of an opening, the approximate amount of water flowing can be determined.

Pressure Reducing Valve a valve in a fire protection system that is designed to limit the downstream water pressure during flowing and non-flowing conditions.

Pump a device that imparts force into a fluid, usually to overcome gravity, friction or containment, for the purpose of transport or pressure application.

Quality Assurance all those planned and systematic actions necessary to provide adequate confidence that a product or service will satisfy given requirements for quality.

Quality Control the operational techniques and activities that are used to fulfill requirements for quality.

Rated Pump Capacity flow capacity of a pump at its rated pressure and speed.

Rate Pump Pressure pressure developed by a pump when operating at its rated capacity.

Relief Valve a valve provided at the discharge of a fire pump which is used to limit the pressure in the fire protection system under abnormal conditions.

Risk Area defined area where a fire hazard is expected to be contained (due to spacing, fire barriers or breaks, etc.) for the purposes of estimating the required fire flow requirements of a facility in order to design the size of firewater pumping capacity.

Sequence Pump Starting a sequence starting device that allows a few seconds between the starting of motors or engines that drive a pump.

Service Factor a multiplier which is applied to a rated horsepower of an alternating current (AC) motor, and indicates a permissible horsepower loading that can be carried at the rated voltage, frequency and temperature. For example, a service factor of 1.1 implies that a motor is permitted to be overloaded 1.1 times the rated horsepower without insulation breakdown or otherwise significantly reducing its service life.

Shutoff or Churn Pressure the net pressure developed by a pump at its rated speed with no flow occurring.

Single Point Failure (SPF) a location in a system so that if failure occurs it will cause the entire system to fail, because backup or alternative measures to accomplish the task are not available.

Single Stage Pump a pump where the total head is provided by the use of one impeller.

Splash Shield A metal shield between the fire pump and the electric motor driver to prevent water leaking from the fire pump and damaging the electric motor.

Static Water Level the level of water suction source for a pump, when the pump in not in operation.

Stuffing Box Packing an arrangement of rings of packing—normally consisting of a lantern ring for the injection of a lubricating or flushing liquid, and a gland to hold the packing and maintain the desired compression to maintain the seal. The function of the packing is to control leakage at the point of the pump shaft, but is not intended to eliminate it completely. The packing is lubricated by the pumped liquid, for fire pumps this is water. The lantern ring is supplied for situations where the stuffing box pressure exists below atmospheric pressure, to inject lubrication into the stuffing box by the use of a bypass line from the pump discharge to the lantern ring connection.

Submersible Pump a multi-stage, usually turbine type, pump attached to a totally sealed motor drive that is submerged and suspended in a well, pit or reservoir with power cables and discharge piping attached.

Suction Pit a defined area enclosed by open grates and screens filled with water from an open body of water, such as a river, reservoir or pond, which is used as a fire pump water suction source.

Surge see Water Hammer.

Total Discharge Head the gage reading at the discharge flange of the pump that is referenced to the centerline of the pump, plus the velocity head at the point of the pump pressure gage attachment.

Total Head the mathematical difference between the total discharge head and total suction head. Where suction head exists, the total head equates to the total discharge head minus total suction head. Where a suction lift occurs, total head equals total discharge head plus total suction lift.

Total Suction Head the condition when the suction pressure is above atmospheric. The total suction head is the mathematical total of the gage reading at the pump suction nozzle flange that is referenced to the centerline of the pump, and the velocity head at the point of pump pressure gage attachment. It may also be called positive suction pressure.

Total Suction Lift the condition when suction pressure exists below atmospheric. The total suction lift is the mathematical total of the gage reading at the suction nozzle flange of the pump that is referenced to the centerline of the pump, and the velocity head at the point of pump pressure gage attachment.

Torsional Vibrational Analysis (TVA) a study of the torsional vibrational frequencies associated with the rotational elements of a pump, particularly associated with long vertical shaft driven pumps.

Up and Downthrust the net vertical forces acting on a vertical rotor. They are the result of hydraulic pressures acting upon the rotor components, momentum forces associated with liquid flow, and dead mass of the rotor parts. The net force generally varies with the capacity setting of the pump and can change direction.

Vapor Pressure the relative measure of fluid's volatility. Water has a vapor pressure of 14.7 at a temperature of 100 °C (212 °F).

Variable Speed Pressure Limiting Control a control system for limiting the discharge pressure produced by a fire pump which is achieved by reducing the pump driver speed.

Velocity Head the energy of fluid due to its bulk motion and is expressed as $h = v^2/2g$, where g = acceleration due to gravity, v = velocity in the pipe for the medium under consideration.

Vertical In-line Pump a pump where the driver is supported exclusively by the pump and the suction and discharge connections have a common horizontal centerline that intersects the shaft axis.

Vertical Shaft Turbine Pump a centrifugal pump submerged in the liquid being pumped, with one or more impellers discharging into one or more bowls and a vertical eductor or column pipe used to connect the bowls to the discharge head on which the pump driver is mounted.

Viscosity the property of fluid that enables it to resist internal flow. Internal friction of a fluid due to molecular cohesion.

Vortex Plate a device (usually a steel plate) provided at or around the intake of a pump suction bell to prevent the formation of vortices.

Water Hammer an increase or dynamic change in pressure produced as a result of the kinetic energy of the moving mass of liquid being transformed into pressure energy which results in an excessive pressure rise.

Index

D

E

F

Q

R

S

T

FSC
www.fsc.org
MIX
Papier | Fördert gute Waldnutzung
FSC® C083411

Kolibri 360 GmbH
Ferdinand-Jühlke-Straße 7
99095 Erfurt, Deutschland
produktsicherheit@kolibri360.de